CLINICAL MR SPECTROSCOPY

CLINICAL MR SPECTROSCOPY
First Principles

Nouha Salibi, PhD
Mark A. Brown, PhD

Siemens Medical Systems, Inc.

WILEY-LISS

A JOHN WILEY & SONS, INC., PUBLICATION

New York • Chichester • Weinheim • Brisbane • Singapore • Toronto

Library of Congress Cataloging-in-Publication Data

Salibi, Nouha Mikhael.
    Clinical MR spectroscopy : first principles / Nouha Salibi, Mark A. Brown.
        p. cm.
    Includes index.
    ISBN 0-471-18280-X (pbk. : alk. paper)
    1. Nuclear magnetic resonance spectroscopy.   I. Brown, Mark A., 1955–   .   II. Title.
    [DNLM: 1. Nuclear Magnetic Resonance. QU 25 S165c 1998]
RC78.7.N83S25   1998
616.07'548—dc21
DNLM/DLC
for Library of Congress   97-13903
                                                                     CIP

Printed in the United States of America
10 9 8 7 6 5 4 3 2 1

■■■■ CONTENTS

Since its discovery over 50 years ago, the field of magnetic resonance (MR) spectroscopy has become an invaluable means for examining atoms and molecules in a nondestructive manner. In a clinical setting, this method has evolved into MR imaging, a technique for the spatial localization of tissue water and fat. However, one of the most powerful aspects of MR spectroscopy, the ability to perform chemical identification, is lost in the imaging process. In recent years, improvements in hardware and software have made it possible to merge the two techniques. In vivo MR spectroscopy combines the localization features of imaging with the ability for chemical analysis to provide a noninvasive method for the study of biochemical processes in a patient.

Our goal in writing this book is to provide an introduction to the basic principles of MR as applied to clinical spectroscopic studies. We have assumed that most of our readers are familiar with some of the basic principles of MR imaging, but it is not a prerequisite for understanding the text. Readers familiar with the principles of MR imaging will find many of the fundamental principles of MR spectroscopy to be the same. We do not intend this book to be an exhaustive treatise on the theory of MR spectroscopy nor a complete reference guide for its clinical applications. Instead, we have attempted to present those aspects of MR spectroscopy that are most pertinent when applied in a clinical setting and examples illustrating these principles.

This book is organized into six chapters. Chapter 1 provides a general perspective on the place of clinical MR spectroscopy in the field of MR. The basic principles of MR, which are pertinent for clinical spectroscopic studies, are introduced in Chapter 2. Chapter 3 compares the hardware features that are important for MR spectroscopy in contrast to MR imaging. Some of the more common techniques for spatial localization and data analysis are presented in Chapters 4 and 5. Finally, a few examples are presented of the types of clinical questions that can be examined using MR spectroscopy.

We wish to thank Siemens Medical Systems for their support of this project. We also thank Dr. June S. Taylor for her critical evaluation of the manuscript and constructive suggestions regarding its content. We thank the following authors for providing data prior to publication or for permission to reproduce previously published spectra: Dr. Barbara Holshouser for Figures

6.7b to 6.11, Dr. Thian Ng for Figures 6.12b to 6.14, Dr. Jane Park for Figure 6.24, Dr. Anthony Ribeiro for Figures 2.15, 2.17, 2.19 and 2.32, Dr. June Taylor for Figure 6.19 and Dr. Zhiyue Wang for Figures 6.15–6.18 and 6.20–6.22. Finally, we thank Bill Leon for his assistance and the use of the MR system for some of the figures in Chapter 6.

N. S.

M.A.B.

CHAPTER 1

Introduction

One of the most influential methods developed for the analysis of molecular structure in the last 50 years is magnetic resonance (MR). Magnetic resonance is a technique for probing atoms and molecules based upon their interaction with an external magnetic field. The power in the approach lies both in its nondestructive nature and in its sensitivity to the molecular environment of an individual atom. This dual nature to MR has been exploited in the development of five major applications (Fig. 1.1). To the medical community, the most familiar example of this is magnetic resonance imaging (MRI).[1] Magnetic resonance imaging produces images of the internal organs of a patient without the use of ionizing radiation. The MR signals from the water and fat hydrogen atoms are mapped according to their location within the patient. Since its development nearly 25 years ago, MRI has proven invaluable for the visualization of normal tissue as well as the diagnosis of soft tissue diseases.

The other four applications of MR are based on the technique known as nuclear magnetic resonance (NMR) spectroscopy or simply MR spectroscopy (MRS).[2,3] In the years since its discovery, NMR has become one of the major techniques used in chemistry and physics laboratories for the analysis of molecular interactions and for the identification of chemical compounds. Its ability to probe the chemical environment surrounding atomic nuclei has made it a valuable adjunct to other structure determination methods. Studies can be performed using samples of a few milligrams of material, either as solids or in solution. The fundamental principles described in Chapter 2 are common to all MR spectroscopic techniques and were originally derived to explain MR spectral features from nonbiological samples.

Biological applications of MRS have developed in three directions. The most common direction has been in the analysis and identification of protein and macromolecular structure and conformation using high-resolution NMR.[4] The methodology and equipment necessary for these studies are the same as for nonbiological NMR described previously. One class of measurement techniques used for studying biological samples, known as two-dimensional (2D) techniques, are very similar to the spatial localization techniques

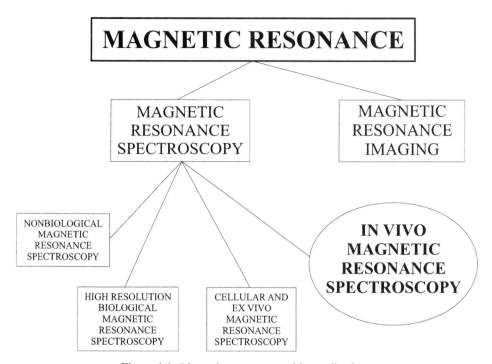

Figure 1.1. Magnetic resonance and its applications.

used in MRI today, and provided the foundation for their development.[5]

The two examples of MRS above (high-resolution nonbiological or biological) are important and powerful methods for understanding molecular interactions but generally are restricted to the study of single molecules or well-characterized systems. For this reason, they require pure samples, that is, samples containing only the molecules under observation with no impurities. This requirement is because the MR signals from the impurities cannot be separated from those of the sample. High magnetic field strengths are typically used (4–14 T) to provide added sensitivity and improved spectral resolution. In contrast, the other two in vivo techniques examine heterogeneous samples: cells, tissues, or entire organ systems. The spectral resolution of these techniques is significantly poorer than for the other techniques, due to the tremendous sample heterogeneity and to the lower magnetic fields typically used for these studies (1–9T). In addition, presently, the more complex techniques such as the 2D approaches, are not often used on these samples and will not be discussed in this book.

The fourth MR method examines multicellular samples or even entire organs ex vivo.[6] These studies usually observe nuclei other than ^{1}H, usually

^{13}C or ^{31}P. While hydrogen nuclei are the most abundant in biological systems, the spectral range of ^{1}H is very narrow and signals from undesired molecules are common. The most common studies of this class acquire multiple spectra at different times and examine metabolic activity or other kinetic processes. Often these studies use enriched isotopes to increase the signal from a known compound above the background and follow its changes due to metabolic reactions. Important insights into cellular biochemistry have been obtained by this method.

The fifth application of MR and the focus of this book is the use of in vivo spectroscopic techniques in a clinical environment. The typical studies compare spectra of pathologic or abnormal tissue and normal tissue. This comparison may be used for diagnosis of pathology or for monitoring of therapeutic treatments. The key to this method is the ability to localize the MR signals to a specific volume of tissue. The localization techniques used here are similar to those used in MRI. However, MRS spectra are acquired from relatively large volumes of tissue (>1 cm^3), because the detected signals are from molecules 1000–100,000 times less concentrated than tissue water in the body.

We have attempted to limit the descriptions of MRS in this book to those aspects that are necessary for understanding the typical MR spectrum. In particular, we have selected those topics that are relevant for understanding MRS studies as performed in a clinical MRI system. Initial applications of MRS techniques to patients have been limited due to inadequate field homogeneity available with whole body magnets, poor coil sensitivity, and low concentration of the metabolites of interest. Improvements in hardware and measurement techniques that evolved for MRI in recent years have made in vivo spectroscopic studies practical. In many cases, spectra with acceptable signal/noise (S/N) ratio and spatial resolution can be obtained with only minimal increase in the total patient examination time. Software improvements allow the data collection and postprocessing to be performed in a streamlined manner. These have enabled MRS to become a feasible and valuable technique to include in a typical patient examination.

REFERENCES

1. R. R. Edelman, J. R. Hesselink, Eds., *Clinical Magnetic Resonance Imaging*, Saunders, Philadelphia, 1990.
2. H. Günther, *NMR Spectroscopy: Basic Principles, Concepts, and Applications in Chemistry*, 2nd ed., Wiley, New York, 1995.
3. D. Shaw, *Fourier Transform N.M.R. Spectroscopy*, 2nd ed., Elsevier Science, Amsterdam, The Netherlands, 1984.

4. R. Ernst, G. Bodenhausen, and A. Wokaun, *Principles of Nuclear Magnetic Resonance Spectroscopy in One and Two Dimensions*, Oxford University Press, Oxford, UK, 1987.

5. W. Brey, Ed., *Pulse Methods in 1D and 2D Liquid Phase-NMR*, Academic, San Diego, CA, 1988.

6. A. I. Scott and R. L. Baxter, Applications of ^{13}C to metabolic studies. *Ann. Rev. Biophys. Bioeng.* **10**, 151–174, 1981.

CHAPTER 2

Concepts of Magnetic Resonance Spectroscopy

2.1. PROPERTIES OF THE ATOM

Magnetic resonance originates from the interaction between an atom and an external magnetic field.[1] Atoms are the constituent components of all matter. The property of the atom necessary for MR is the nuclear spin or spin angular momentum. Every element found in nature, except for cerium and argon, has at least one isotope with nuclear spin; therefore, nearly every element can be studied using MR. The basic concepts of energy absorption, chemical shift, and relaxation are common to all such nuclei, while the specific details are unique to each nucleus.

The atomic structure provides the basis for describing many of the properties of atoms. Within an atom are three component parts in varying amounts: protons, which have a positive charge; electrons, which have a negative charge; and neutrons, which have no charge (neutral). The protons and neutrons form the nucleus or core to the atom while the electrons surround the nucleus. Three characteristic properties of the atom are of fundamental importance. The atomic number is the number of protons in the nucleus. It is the primary index used to identify elements. All atoms of an element have the same atomic number. The atomic weight is the total number of protons and neutrons. The protons and neutrons have almost equal mass and are 1800 times heavier than the electrons so that the nucleus contains the bulk of the atomic mass. Atoms of the same element (atomic number) with different atomic weights are called isotopes. For elements with multiple isotopes, the relative amounts of each isotope or the natural abundance is an important consideration regarding the ability to examine the nucleus using MR techniques. For example, hydrogen has three isotopes that occur in nature: ^1H or protium, ^2H or deuterium, and ^3H or tritium. All three isotopes have one proton in the nucleus, but have zero, one, or two neutrons, respectively. The protium isotope is found in 99.9% of all hydrogen atoms, which makes it a natural choice for MR studies of hydrogen such as spectroscopy or imaging.

The deuterium isotope is approximately 0.1% of all hydrogen while tritium is significantly less than 0.1%.

The third property used to describe atoms is the nuclear spin or intrinsic nuclear spin angular momentum. The nuclear spin describes the intrinsic motion of the nucleus. A complete description of the nuclear spin and its properties requires the use of a mathematical model known as quantum mechanics. For most purposes in clinical spectroscopy, a few key features of quantum mechanics are sufficient for describing the behavior of nuclei. In addition, for biological systems, a so-called "high temperature" approximation validates the use of classical mechanics for most situations.

Nuclear spin is important because it is a requirement for MR. One important characteristic of nuclear spin is that it is quantized or has certain discrete values. These values depend on the specific nature of the nucleus. Three classes of values for the nuclear spin are found:

1. Zero or no spin. A spin of zero is found for nuclei with an even atomic weight and atomic number (even numbers of protons and neutrons). Nuclei with no spin cannot be examined using MR.

2. Integer or whole number. An integer spin (1, 2, 3, . . .) is found for nuclei with an even atomic weight and an odd atomic number (odd numbers of protons and neutrons).

3. Half-integer. A half-integer spin ($\frac{1}{2}$, $\frac{3}{2}$, $\frac{5}{2}$, . . .) is found for nuclei with an odd atomic weight (even number of protons and odd number of neutrons, or vice versa). For clinical spectroscopy, this is the most common class of spins encountered (^{1}H, ^{19}F, ^{23}Na, ^{31}P). Neutrons and electrons also have a spin of $\frac{1}{2}$.

An important consequence of nuclear spin is that it has associated with it a magnetic field. A useful analogy for a nucleus with spin is a bar magnet (Fig. 2.1). A bar magnet has a north and south pole, or more precisely, a magnitude and orientation to the magnetic field can be defined. A nucleus with spin can be viewed as a vector or magnetic dipole having a definite magnitude and an axis of rotation with a definite orientation. The manipulation of the spin orientation by a radiofrequency (rf) pulse is the basis for the MR measurement.

2.2. PRODUCTION OF NET MAGNETIZATION

Imagine an arbitrary volume of tissue containing a group of identical spins located outside a magnetic field. Each spin can be represented as a vector of

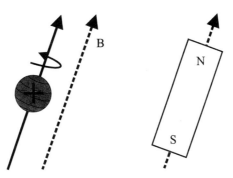

Figure 2.1. A nuclear spin with a positive charge produces a magnetic field that is oriented in the direction of the axis of rotation. This concept is analogous to a bar magnet in which the magnet field is considered to be oriented between the south and the north pole.

equal magnitude.[a] However, the spin vectors for the entire collection of spins within the tissue are randomly oriented in all directions. Performing a vector addition of these spin vectors produces a zero sum; that is, no net magnetization is observed in the tissue (Fig. 2.2).

If the tissue is placed inside a magnetic field B_0, the situation changes. The individual spins will begin to rotate, or precess, about the magnetic field. They will also be tilted slightly away from the axis of the magnetic field, but the axis of rotation will be parallel to B_0. This precession occurs because of the interaction of the magnetic field with the moving positive charge of the nucleus. By convention, B_0 is defined to be oriented in the z direction of a Cartesian coordinate system; the axis of precession will also be the z axis. The motion of each spin can be described by a unique set of x, y (perpendicular to B_0), and z (parallel to B_0) coordinates. The perpendicular, or transverse, coordinates are nonzero and will vary with time as the spin precesses, but the z coordinate is constant with time (Fig. 2.3). The rate or frequency of precession is proportional to the strength of the magnetic field and is expressed by Eq. (2.1), the Larmor equation.

$$\omega_0 = \gamma B_0/2\pi \tag{2.1}$$

where ω_0 is the Larmor frequency in megahertz (MHz), B_0 is the magnetic field strength in tesla (T) that the spin experiences, and γ is a constant for each nucleus in units of reciprocal seconds reciprocal tesla ($s^{-1}T^{-1}$), known as the gyromagnetic ratio. Values for γ and ω_0 at 1.5 T for several nuclei are listed in Table 2.1.

[a]Vector quantities with direction and magnitude are indicated by boldface type while scalar quantities that are magnitude only are indicated by regular type.

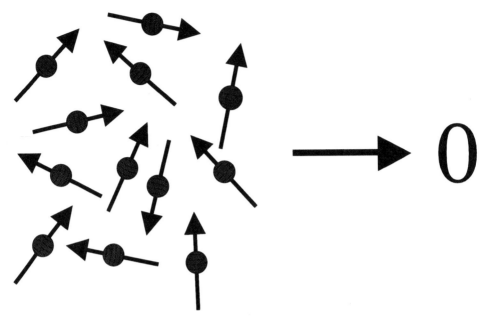

Figure 2.2. Microscopic picture and macroscopic picture of a collection of spins in the absence of an external magnetic field. In the absence of a magnetic field, the spins will have their spin vectors oriented randomly (microscopic picture). The vector sum of these spin vectors will be zero (macroscopic picture).

There are two methods commonly used to express the frequency of a time-varying wave. The cyclical approach analyzes the absolute number of variations or cycles of the wave per unit time, usually seconds [1 cycle s^{-1} = 1 hertz (Hz)]. This method is particularly useful for analyzing plane waves or waves that are two dimensional. The angular or radial approach measures fractions of a circle arc or radians per unit time (rad s^{-1} or simply s^{-1}). The two measures may be interconverted since 1 cycle = 2π radians (Fig. 2.4). Care must be taken to ensure the correct measure of frequency is used. Most theoretical derivations of MR use angular units since the precessional motion of spins is assumed to be circular in one direction. Most practical analyses use cyclical units because the waves produced by the MR hardware are plane waves.[b]

[b]An additional source of confusion is the choice of variables used to represent the frequency in the Larmor equation. Historically, angular frequency is represented by ω (Greek letter omega) while cyclical frequency is represented by ν (Greek letter nu) or f. The Larmor equation is more properly written $\nu = \gamma B_0/2\pi$. Common usage of Eq. (2.1) in imaging derivations uses ω but with the units of hertz. To minimize confusion, we will follow the imaging tradition throughout this book.

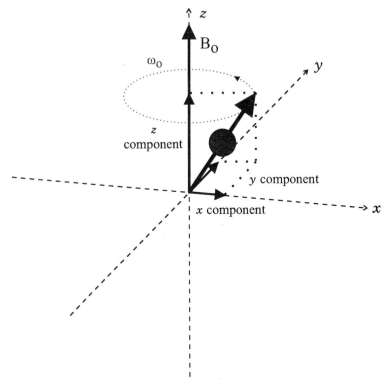

Figure 2.3. Inside a magnetic field, a nucleus with spin precesses or revolves about the magnetic field. The precessional axis is parallel to the main magnetic field **B_0**. The z component of the nuclear spin vector (projection of the spin onto the z axis) is the component of interest since this component does not change in magnitude or direction as the spin precesses. The x and y components vary with time at a frequency ω_0 proportional to B_0 as expressed by the Larmor equation.

The interaction of a spin with the external magnetic field produces an additional effect. Because the axis of precession for the spin is constant with time (time invariant), there will be a nonzero interaction or coupling between the spin and **B_0**, known as the Zeeman interaction. The Zeeman interaction causes the z component of the quantized spin to assume discrete values as well. The possible values depend on the value for the spin, I. There are two selection rules:

1. The total number of possible z components is equal to $2I + 1$, symmetrically divided about 0.
2. The difference in z component values between adjacent orientations is equal to $+1$ or -1.

TABLE 2.1. Nuclear Spin Parameters^a

Isotope	Natural Abundance (%)	I	γ (MHz T^{-1})	ω at 1.5 T (MHz)	Relative Sensitivity
^{1}H	99.985	½	42.5774	63.8646	100.00
^{2}H	0.015	1	6.53896	9.8036	0.94164
^{6}Li	7.5	½	6.26613	9.3990	0.84932
^{7}Li	92.5	3/2	16.5483	24.8217	29.355
^{12}C	98.9	0			
^{13}C	1.10	½	10.7084	16.0621	1.5909
^{14}N	99.63	1	3.07770	4.6164	0.53029
^{19}F	100	½	40.0776	60.1148	83.400
^{23}Na	100	3/2	11.2686	16.9029	9.2697
^{31}P	100	½	17.2514	25.8765	6.6518

^{a}Adapted from Mills.

For a spin ½ nucleus, there are only two possible values for the z component that satisfy these rules: $+½$ and $-½$, commonly called spin up (parallel) and spin down (antiparallel); for a spin 1 nucleus, there are three possible values: $+1$, 0, and -1; for a spin 3/2 nucleus, there are four possible values: $+3/2$, $+½$, $-½$, and $-3/2$. Only these values for the z component are possible (Fig. 2.5).

The Zeeman interaction also affects the energy of a spin. The coupling between the magnetic field and the spin produces a quantization of the energy states or levels for a spin. There will be a total of $2I + 1$ energy

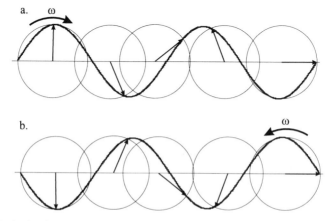

Figure 2.4. A circular wave at a frequency ω can be mapped into a plane wave at the same frequency ω. While the circular wave can be in either direction, the plane wave that is produced is the same in both cases. (*a*) Forward direction circular wave. (*b*) Reverse direction circular wave.

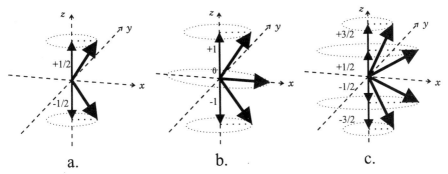

Figure 2.5. Zeeman interaction. In the presence of a magnetic field (z direction, not shown), the x and y components of the spin vector vary with time as the nucleus precesses. The spin vector traverses a circle in the xy plane. The z component has discrete values, depending on the value for the spin, I. (*a*) A spin ½ nucleus has two possible z components ($+½, -½$). (*b*) A spin 1 nucleus has three possible z components ($+1, 0, -1$). (*c*) A spin ³⁄₂ nucleus has four possible z components ($+³⁄₂, +½, -½, -³⁄₂$).

levels, each one corresponding to a spin with a different z component. For most nuclei, the lowest energy state corresponds to the largest positive z component (parallel to **B₀**) and the highest energy state corresponds to the largest negative z component (antiparallel to **B₀**). Most nuclei studied in biological systems using MR techniques have a spin of ½ and have the parallel component ($z = +½$) the lowest in energy. The difference in energy between the energy levels is proportional to B_0 (Fig. 2.6). The frequency

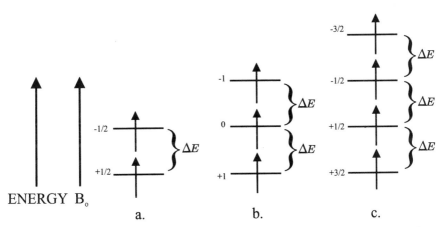

Figure 2.6. Zeeman interaction. In the presence of a magnetic field, the energy states for a spin become unequal. The difference in energy ΔE between any two states is the same and is proportional to B_0. For most spin ½ nuclei, the $z = +½$ level is lowest in energy (*a*); for most spin 1 nuclei, the $z = +1$ level is lowest in energy (*b*); for most spin ³⁄₂ nuclei, the $z = +³⁄₂$ level is lowest in energy (*c*).

corresponding to this energy difference is exactly the Larmor frequency ω_0 of the spin. Energy applied to the sample at this frequency, also known as the resonant frequency, will cause the spin to go from one energy level to another one through a process called resonance absorption. Only energy at this frequency will cause a spin to change energy levels.

For practical reasons, individual spins are not examined in MR; instead, the entire collection of spins inside the volume of tissue is treated as a single unit because this bulk entity is easiest to observe. The number of spins per milliliter (mL) of tissue depends on the specific nucleus involved. For example, there are approximately 10^{24} ^{1}H atoms per mL of tissue, mostly in water molecules. The Zeeman interaction described above occurs for every spin inside the volume of tissue. The entire collection of spins will be distributed between the various possible energy levels. Since these levels are not equal in energy, the distribution will be unequal between the various possible levels, with more spins occupying the lower levels. The distribution is described by the Boltzmann equation:

$$N_i = N_{\text{total}} e^{(-E_i/kT)} \qquad (2.2)$$

where N_i is the number of spins in state i (e.g., spin up or spin down for spin $\frac{1}{2}$ nuclei), E_i is the energy of state i measured in Joules (J), N_{total} is the total number of spins in the volume, k is a constant known as Boltzmann's constant (1.381×10^{-23} J K^{-1}), and T is the absolute temperature of the sample in degrees kelvin (K). A more useful form of this equation is obtained by comparing the relative number of spins between two states i and j, such as for spin $\frac{1}{2}$ nuclei:

$$N_i/N_j = e^{(-\Delta E/kT)} \qquad (2.3)$$

where ΔE is the difference in energy between the two states. The Boltzmann equation indicates that the largest number of spins will be in the states of lower energy with gradually reduced numbers in the states of higher energy (Fig. 2.7). The difference in the number of spins in each state, and thus the amount of polarization induced in the sample by the magnetic field, is very small. For ^{1}H atoms at room temperature, this excess of spins in the lower energy state is approximately $1:10^6$. This very weak spin polarization results in MR signals that are very low in amplitude and require care to ensure proper detection.

The quantum mechanical description of spin interaction with the magnetic field summarized previously is necessary for the explanation of many experiments in MR spectroscopy, but it is not particularly intuitive. An alternate perspective, the classical approach, can be used to describe many features of

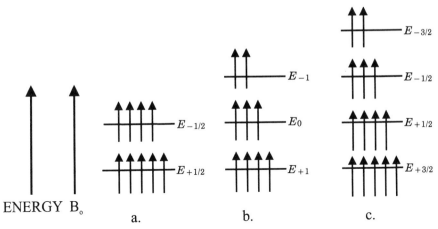

Figure 2.7. Boltzmann diagram. The number of spins in each energy level decreases as the energy of the level increases. (*a*) Spin ½ case, (*b*) spin 1 case, and (*c*) spin ¾ case.

MR spectroscopy. For spin ½ nuclei, this model is particularly useful and simple. As mentioned above, there will be a slight excess of spins in the lower energy, spin-up state compared to the higher energy, spin-down state. This unequal number of spins in each energy level means that the vector sum of spins will be nonzero and will point parallel to the magnetic field. In other words, the tissue will become magnetized in the presence of $\mathbf{B_0}$ with a value $\mathbf{M_0}$, known as the net magnetization. Its magnitude is proportional to B_0:

$$\mathbf{M_0} = \chi \mathbf{B_0} \tag{2.4}$$

where χ is the nuclear magnetic susceptibility. The orientation of $\mathbf{M_0}$ will be in the same direction as $\mathbf{B_0}$ and will be constant with respect to time (Fig. 2.8). This arrangement with $\mathbf{M_0}$ aligned along the magnetic field with no transverse component is the normal, or equilibrium, configuration for the spins. It is the lowest energy configuration and the arrangement to which the spins naturally try to return following any perturbation, such as energy absorption. The induced magnetization, $\mathbf{M_0}$, is the source of signal for all MR experiments. Consequently, all other things being equal, the greater the field strength, the greater the value of $\mathbf{M_0}$ and the greater the MR signal.

Another way to visualize this net magnetization is to recall that the individual spins precess about the magnetic field. When energy absorption is included, this precessional motion continues but becomes more complicated to describe. A useful concept to simplify the description is called a rotating frame of reference or rotating coordinate system. In the rotating frame, the coordinate system rotates about one axis while the other two axes vary with

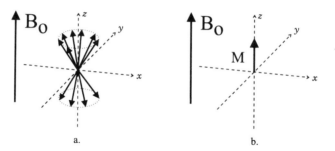

Figure 2.8. Microscopic (*a*) and macroscopic (*b*) picture of a collection of protons in the presence of an external magnetic field. Each proton precesses about the magnetic field. If a rotating frame of reference is used with a rotation rate of ω_0, then the collection of protons appears stationary. While the *z* components are one of two values (one positive and one negative), the *x* and *y* components can be any value, positive or negative. The protons will appear to track along two "cones," one with positive *z* component and one with negative *z* component. Because there are more protons in the upper cone (parallel to $\mathbf{B_0}$), there will be a nonzero vector sum $\mathbf{M_0}$, the net magnetization. It will be of constant magnitude and parallel to $\mathbf{B_0}$.

time. By choosing a suitable axis and rate of rotation for the coordinate system, the moving objects appear stationary.

For MR experiments, a convenient rotating frame uses the *z* axis, parallel to $\mathbf{B_0}$, as the axis of rotation while the *x* and *y* axes rotate at the Larmor frequency ω_0. When viewed in this fashion, the precessing spin will appear stationary in space with a fixed set of *x*, *y*, and *z* coordinates. If the entire collection of spins in the tissue volume is examined, a complete range of *x* and *y* coordinates will be found, both positive and negative, but only two *z* coordinates (Fig. 2.8). There will be an equal number of positive and negative *x* and *y* values, but a slight excess of positive *z* values, due to the Zeeman interaction as described above. If a vector sum is performed on this collection of protons, the *x* and *y* components sum to zero but a nonzero, positive *z* sum will be left that is the net magnetization $\mathbf{M_0}$. In addition, since the *z* axis is the axis of rotation, $\mathbf{M_0}$ does not vary with time. Regardless of whether a stationary or fixed coordinate system is used, it is of fixed amplitude and is parallel to the main magnetic field. For all subsequent discussions in this book, a rotating frame of reference with the rotation axis parallel to $\mathbf{B_0}$ will be used when describing the motion of the spins.

2.3. RESONANCE ABSORPTION

The MR experiment, in its simplest form, can be considered to be a stimulated emission phenomenon. In any MR measurement, energy is applied to the sample or patient that will be absorbed. A short time later, this energy will be emitted at which time it is detected and processed. The nature of the absorption and emission processes are beyond the scope of this discussion and do not provide any particular insight. However, the results following absorption and emission can be analyzed in terms of both the molecular and macroscopic models for a collection of spins to explain the observed results. Because spin ½ nuclei are the most common nuclei studied using in vivo spectroscopy, the subsequent description of resonance absorption will focus on them.

In Section 2.2, the Zeeman interaction and its effect on a collection of spins were described. The difference in energy between any two of the spin states is proportional to the magnetic field that the spins experience. If a pulse or short burst of energy that matches this difference is applied to the sample, a spin in the lower energy state will be excited to the upper energy state (Fig. 2.9). Only energy corresponding to this difference will produce this excitation. Following the pulse, the spins will reemit the energy at the same frequency and return to the lower energy state. The particular frequency that is absorbed is proportional to the magnetic field $\mathbf{B_0}$ and the equation describing it is the Larmor equation, Eq. (2.1). This process is known as resonance absorption and the frequency of the absorbed energy is known as the resonant frequency for the spins.

It is more useful to discuss the resonance condition by examining the effect of energy absorption on the net magnetization $\mathbf{M_0}$ rather than on an individual spin. When applied to a large collection of spins such as water protons in a volume of tissue, an rf pulse produces a significant amount of both absorption (spins in lower state going to higher state) and emission (spins in higher state going to lower state). However, because there are more

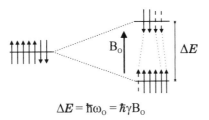

$$\Delta E = \hbar \omega_0 = \hbar \gamma B_0$$

Figure 2.9. Zeeman diagram. The difference in energy between the two configurations (spin up and spin down) is proportional to the magnetic field strength B_0 and the corresponding precessional frequency ω_0. When energy of this frequency is applied, a spin from the lower energy state is excited to the upper energy state. Also, a spin from the upper energy state is stimulated to give up its energy and go to the lower energy state. Because more spins are in the lower energy state, there is a net absorption of energy by the spins in the sample. $\hbar$ is Planck's constant h (6.626×10^{-34} J s) divided by 2π.

spins in the lower energy level, there will be more absorption rather than emission, resulting in a net absorption of energy by the tissue. The energy is applied as a radiofrequency (rf) pulse with a central frequency ω_1 and a magnetic field oriented perpendicular to $\mathbf{B_0}$ (indicated by a magnetic field $\mathbf{B_1}$) (Fig. 2.10). This orientation difference between $\mathbf{B_0}$ and $\mathbf{B_1}$ allows the rf pulse and $\mathbf{M_0}$ to couple so that energy can be transferred to the spins. This transfer is most efficient if ω_1 is equal to ω_0. Absorption of the rf energy of frequency ω_0 causes $\mathbf{M_0}$ to rotate away from its equilibrium orientation. If the rf pulse is applied long enough and at a high enough amplitude, the absorbed energy will cause $\mathbf{M_0}$ to rotate entirely into the transverse plane. Such a pulse is known as a 90° pulse. The direction of rotation of $\mathbf{M_0}$ is perpendicular to both $\mathbf{B_0}$ and $\mathbf{B_1}$. When viewed in a rotating frame at frequency ω_0, the motion of $\mathbf{M_0}$ is a simple vector rotation.

When the rf pulse is turned off, the spins immediately begin to realign themselves and return to their original equilibrium orientation. They will emit energy at frequency ω_0 as they do. If a loop of wire (receiver coil) is placed with its axis parallel to the transverse plane, the spins will induce a voltage in the wire during their precession. This voltage will decay with time as more of the spins lose their absorbed energy through a process known as relaxation (see below). The induced voltage is the MR signal and is known as the free induction decay (FID) [Fig. 2.11(a)]. Only the portion of the magnetization that is perpendicular to the external magnetic field will induce a voltage in the receiver coil. The FID is measured and stored for later postprocessing. The magnitude of the FID depends on the value of $\mathbf{M_0}$ immediately prior to the 90° pulse.

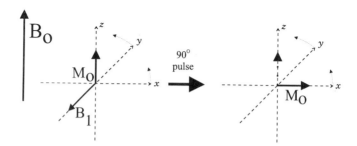

Figure 2.10. Effect of a 90° rf pulse in a frame of reference rotating at ω_0. The rf pulse broadcast at frequency ω_1 can be treated as an additional magnetic field $\mathbf{B_1}$ oriented perpendicular to $\mathbf{B_0}$. When the pulse is applied at the appropriate frequency ($\omega_1 = \omega_0$), the protons absorb it and $\mathbf{M_0}$ rotates into the transverse plane. The direction of rotation is perpendicular to both $\mathbf{B_0}$ and $\mathbf{B_1}$.

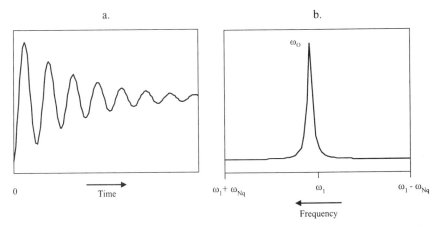

Figure 2.11. (*a*) Free induction decay, real part. The response of the net magnetization $\mathbf{M_0}$ to an rf pulse is known as the FID. It is proportional to the amount of transverse magnetization generated by the pulse. The FID is maximized when using a 90° excitation pulse. (*b*) Magnitude Fourier transformation of (*a*). The Fourier transformation is used to convert the digital version of the MR signal (FID) from a function of time to a function of frequency. Signals measured with a quadrature detector are displayed with the transmitter (reference) frequency ω_1 in the middle of the display.

2.4. SIGNAL ANALYSIS AND THE MR SPECTRUM

If all the spins experience the same magnetic field $\mathbf{B_0}$, then only one frequency would be contained within the FID. Due to the interactions described in later sections, each type of spin will experience its own local magnetic field. There are many such magnetic fields throughout the magnet, producing many MR signals at many frequencies following the rf pulse. These signals are superimposed so that the FID consists of many frequencies as a function of time. Because the frequencies are relatively high, the FID that is usually analyzed is a demodulated or difference signal from the original transmitter (see Section 3.4); that is, the MR signal that is examined is in a rotating frame relative to the frequency of the transmitter pulse. Most MR receivers perform two such demodulations relative to the transmitter frequency, one with the reference signal in phase with the original transmitter pulse and one 90° shifted in phase. The phase shifting between the two receivers necessitates the use of complex analysis for processing the signal. This type of receiver is known as a quadrature receiver, and the two signals are called the real and imaginary channels.

Time-domain analysis of the MR signal can be useful in many instances, particularly if the signal contains only one frequency. However, multicomponent signals are easier to examine in terms of frequency rather than of time. The conversion of the signal amplitudes from a function of time to a function of frequency is accomplished using a mathematical operation called the Fourier transformation. In the frequency presentation or frequency domain spectrum, the MR signal is mapped according to its frequency relative to the transmitter frequency ω_1. For systems using quadrature detectors, ω_1 is centered in the display with frequencies higher and lower than ω_1 located to the left and right, respectively [Fig. 2.11(b)]. The frequency domain allows a simple way to examine the magnetic environment that a spin experiences.

While allowing a simple presentation of the frequencies generated by a collection of spins, the Fourier transformation produces a complication to the resulting spectrum. The response of a spin to an rf pulse can be divided into two parts: one that is in phase with the transmitter, known as dispersion, and one that is 90° out of phase, known as absorption [Fig. 2.12(a)]. These correspond to the imaginary and real parts of the frequency spectrum, respectively (so named for the respective components used in their mathemati-

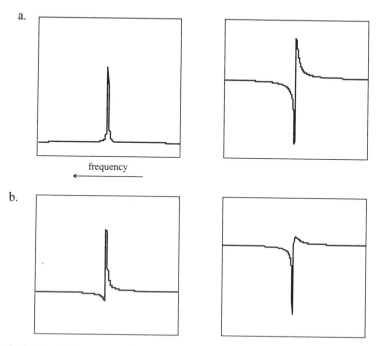

Figure 2.12. The ^{1}H spectrum of water. (a) Phase corrected. Left: Absorption mode and right: dispersion mode. (b) Phase uncorrected. Spectra are mixtures of absorption and dispersion modes.

cal description). Because signal detection is not instantaneous following the excitation pulse, the Fourier transformation mixes these two responses, so that the resulting spectrum is a linear combination of both absorption and dispersion modes [Fig. 2.12(*b*)]. While the signal amplitudes of the absorption and dispersion spectra can be individually related to the number of spins producing the signal, the magnitude spectrum, which is produced from the combined signals, cannot. For this reason, a process known as phase correction is normally performed. Phase correction allows the two spectra to be separated so that further analysis may be performed, usually on the absorption spectrum.

There are three quantities of the MR absorption spectrum that are of particular interest: the integrated area, position (resonant frequency), and width [full width at half-maximum height (fhwm)], of each resonance peak. The peak area is proportional to the number of spins producing the signal. Because there are several factors that determine the signal that are constant but difficult to measure, the typical approach is to analyze ratios of peak area between resonances within the spectrum. The peak position is used to identify the type of spin and its particular molecular environment. The interactions responsible for the position of a particular peak are usually constant with time (time independent), but in certain instances may be time dependent.

Interpretation of the peak width is most accurately performed by determination of the shape of the spectral peak. Most MR spectra of tissues are combinations of two line shapes. One is Lorentzian, the shape observed when the spins are in a single molecular environment and are rapidly tumbling, usually a small molecule [Fig. 2.13(*a*)]. Lorentzian line shapes have the property that the line width [half-width at half-maximum height (hwhm)] is

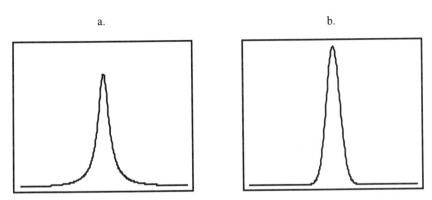

a.   b.

Figure 2.13. Absorption mode line shapes. Curves are plotted assuming equal hwhm, normalized to unit area: (*a*) Lorentzian line shape and (*b*) Gaussian line shape.

inversely proportional to the total transverse relaxation time, T2* (see below). Lorentzian line shapes also have very broad bases or "wings," which contribute significantly to the overall integrated area for the resonance. The other line shape is Gaussian, and is observed when spins in multiple environments that are very similar contribute to the resonance [Fig. 2.13(b)]. Line shape analysis is often used to identify whether the signal comes from one type of spin or from multiple types. The formulas and relationships for both line shapes and their corresponding time-domain functions are listed in Table 2.2.

2.5. TIME-INDEPENDENT INTERACTIONS

The model, described in section 2.2, of the interaction of spins with a magnetic field and the Larmor equation is sufficient to analyze isolated nuclei in a magnetic field. For atoms within molecules, this simple description is insufficient to explain the MR spectra. Consideration of the molecular structure is necessary. The ensuing spectral modifications may be categorized according to their time-varying nature. Two interactions, chemical shielding and spin coupling, are constant and do not change with time during the measurement; they affect primarily the resonant frequency of the spins. These are of primary importance in identification of individual molecules within the spectrum. Two other interactions, relaxation and chemical exchange, are time-dependent phenomena and affect both the resonant frequency and signal amplitude of the spins. In all cases, we will make the assumption that these interactions are small perturbations to the Zeeman interaction and the primary resonance absorption process described above. For most spin systems in clinical MR spectroscopy, this so-called "high-field approximation" is true and provides significant simplification of the experimental situation.

2.5.1. Chemical Shift

As mentioned in section 2.1, atoms are surrounded by electrons, which also possess spin. In molecules, these electrons generally form bonds between atoms. The electrons are not located entirely near one atom or the other in a bond, but are distributed or shared between the two atoms, forming a "cloud." An exact description of this charge distribution requires the use of quantum mechanics. In general, however, the shape of the electronic charge distribution depends on the particular nuclei involved in the molecular bond. For example, oxygen–hydrogen bonds have most of the electron density on the oxygen atom while carbon–hydrogen bonds have a more equal electronic distribution.

TABLE 2.2. Line Shape Functions[a]

	Lorentzian	Gaussian
Time Domain	$\exp(-t/T2^*)(\cos \omega_o t + i \sin \omega_o t)$	$\exp(-t^2/T_d^2)(\cos \omega_o t + i \sin \omega_o t)$
Frequency domain line width (hwhm), Hz	$\dfrac{1}{2\pi T2^*}$	$\dfrac{\sqrt{\ln 2}}{\pi T_d}$
Frequency domain absorption	$\dfrac{\text{hwhm}/\pi}{(\text{hwhm})^2 + (\omega - \omega_o)^2}$	$\dfrac{\sqrt{\ln 2} * \exp\{-[(\omega - \omega_o)/\text{hwhm}]^2\}}{\sqrt{\pi}}$
Frequency domain dispersion	$\dfrac{i(\omega - \omega_o)/\pi}{(\text{hwhm})^2 + (\omega - \omega_o)^2}$	—
Frequency domain magnitude	$\dfrac{1}{\pi * [(\text{hwhm})^2 + (\omega - \omega_o)^2]^{1/2}}$	—

[a] Adapted from Shaw.[8] Frequency domain absorption line shapes are normalized to unit area.

In the presence of an external magnetic field, the electrons will also interact-with the field. The negative charge of the electronic spin generates a magnet-ic field that opposes or shields the nucleus from the external magnetic field. This is known as chemical shielding. The extent of shielding of the nucleus depends on the local electron density and thus the particular atoms to which the nucleus is bonded. The incorporation of chemical shielding in the de-scription of MR requires modification of the total magnetic field to include two components, the main magnetic field and the shielding term (Eq. 2.5):

$$\mathbf{B}_i = \mathbf{B}_0(1 - \sigma_i) \tag{2.5}$$

where σ_i is the chemical shielding term for nucleus i. In general, chemical shielding is very small compared to the main magnetic field and is usually considered to be a perturbation to the main field. Chemical shielding is also a tensoral quantity, with orientational dependencies in all three directions (x, y, and z). However, in many experimental situations including biological sys-tems, it is sufficient to substitute the average value σ_i of the principal components of the shielding tensor into Eq. 2.5. This will be the case whenever the spins are in small molecules or in parts of molecules that tumble rapidly enough to average all orientations.

Chemical shielding also affects the Zeeman splitting shown in Figure 2.6 and the Boltzmann energy levels of Eq. 2.2. Each energy level is modified by the shielding since the total magnetic field experienced by a spin is different. However, the effect is slightly different for each of the levels of that spin. For example, the two energy levels for methyl protons in ethanol (spin up and spin down) are affected in opposite ways by the incorporation of shielding of the methyl protons (Fig. 2.14). The resonant frequency as expressed by the Larmor equation (Eq. 2.1) must also be modified:

$$\omega_i = \gamma B_0(1 - \sigma_i) \tag{2.6}$$

where ω_i is the frequency for the ith spin. This modified Larmor equation indicates that the resonant frequency of a spin depends on its molecular environment as well as the external magnetic field that it experiences.

The normal method for display of MR spectra is to plot high frequencies on the left and low frequencies on the right.[c] When displayed full scale, the

---

[c]The reason for this "backward" display is historical. Originally, MR spectroscopy used a continuous rf signal (continuous wave, CW) and varied **B** to achieve resonance, a technique known as field sweep. In this perspective, the plot is from low field values on the left and high field values on the right. Current data processing methods can display spectra in either direction, but the original perspective was retained to preserve consistency with earlier work.

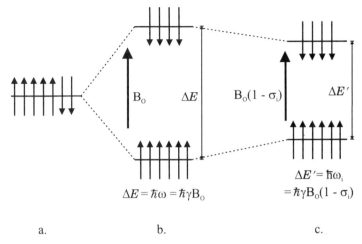

Figure 2.14. Zeeman diagram. (*a*). In the absence of the magnetic field, the spin-up and spin-down orientations are equal in energy. (*b*) In the presence of a magnetic field **B₀** only the difference in energy ΔE is proportional to B_0. (*c*) When the effect of chemical shielding is included, the total magnetic field and energy difference is reduced from (*b*). This shielding term depends on the molecular environment of the spin.

transmitter frequency is in the center and spectral frequencies above and below the transmitter are plotted on the left and right, respectively. While these frequency differences may be plotted in units of Hz, a more convenient scale to use is parts per million (ppm). The ppm scale defines frequency differences for the resonant frequency of the spin of interest relative to a reference frequency:

$$\delta_i = (\omega_i - \omega_{ref})/\omega_{ref} \tag{2.7}$$

Frequency differences expressed as relative differences δ_i are known as chemical shifts and are usually expressed in units of ppm away from the reference frequency. Chemical shifts are particularly useful because they are independent of B_0; in other words, tables of chemical shifts may be used at all field strengths. For example, the resonant frequency of fat protons differs from the resonant frequency of water protons by 150 Hz at 1.0 T and by 225 Hz at 1.5 T, yet this difference is 3.5 ppm in both cases. Table 2.3 lists chemical shifts for ¹H and ³¹P nuclei in some biological molecules. While chemical shifts are generally a nuisance in imaging, causing signal mis-registrations known as chemical shift artifacts, they provide a wealth of information in spectroscopic studies. Chemical shifts enable identification of nuclei or functional groups within molecules, which permits a piecewise determination of the molecular structure. This method provides a valuable

TABLE 2.3. Chemical Shift Values for Selected Resonances[a]

Functional Group/Molecule	Chemical Shift (ppm)
Lactate, methyl group	1.3
N-Acetylaspartate (NAA), methyl group	2.0
Creatine/phosphocreatine, methyl group	3.0
Choline, methyl group	3.2
Choline, methylene group	3.9
Water (H_2O)	4.7
α-ATP	−7.8
β-ATP	−18.2
γ-ATP	−2.7
Phosphocreatine phosphate group	0.0
Inorganic phosphate (P_i)	5.2

[a]The ^1H chemical shifts are defined relative to tetramethylsilane [TMS, $(Me)_4Si$] at 0.0 ppm, and adapted from Frahm et al[9]. The ^{31}P chemical shifts are defined relative to phosphocreatine at 0.0 ppm. [Adapted from Burt et al.[10]]

tool for probing chemical bonding and other interactions. Figure 2.15 illustrates this for the three types of protons found in ethanol.

For in vivo spectra, identification of chemical shifts to chemical structures on specific molecules must be made carefully. On a practical basis, the chemical shielding effects are limited to a few bonds away from the nucleus of interest at the field strengths used (<4 T for whole body systems, <11 T for smaller bore systems). Because larger molecules may have similar functional groups, they may have similar spectra. For example, adenosine triphosphate (ATP) and guanosine triphosphate (GTP) in vivo have indistinguishable ^{31}P spectra since the molecular differences between adenosine and guanosine are several bonds away from the phosphorus nuclei (Fig. 2.16).

2.5.2. Spin Coupling

Another molecular interaction that modifies the resonant frequency of a spin is known as spin coupling. Spins within a molecule interact with each other and will affect the local magnetic field around each nucleus. The most common manifestation of this interaction in biological systems is facilitated by the electrons adjacent to the nuclei. This is known as spin–spin coupling or J coupling, for the variable that has traditionally been used to describe it. Other types of spin coupling interactions are important in the studies of pure solids such as crystals, but their inclusion in the present description adds unnecessary complications and no particular insight. Spin coupling differs

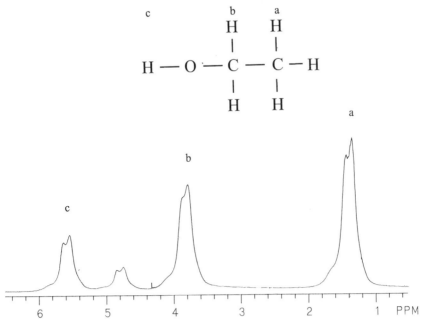

Figure 2.15. ^{1}H spectrum of 95% ethanol in water, moderate spectral resolution. The labeled resonances correspond to the methyl (a), methylene (b), and hydroxyl (c) protons with relative areas of 3:2:1. The unlabeled resonance (4.8 ppm) is from water. Measurement parameters: Frequency: 300 MHz; Number of data points: 32768; Number of acquisitions: 8; TR: 8200 ms; Flip angle: 25°.

from the chemical shift in two important ways: it is independent of magnetic field strength (chemical shift increases with B_0 when measured in Hz) and there is always another spin involved in the coupling.

An understanding of spin coupling may be obtained by considering the ^{1}H MR spectrum of ethanol shown in Figure 2.17. In the presence of a homogeneous magnetic field, the spectrum shows additional resonance signals compared to Figure 2.15. These signals may be explained by considering the molecular structure of ethanol. The methylene group of ethanol [Fig. 2.15(*b*)] has two protons bonded to the carbon atom. Inside a magnetic field, these two protons will be configured in three arrangements: both spins up (total spin 1), both spins down (total spin -1), or one spin up and one spin down (total spin 0) [Fig. 2.18(*b*)]. The last arrangement is twice as likely as the other two arrangements. A proton on an adjacent atom such as a methyl proton [Fig. 2.15(*a*)] will "sense" one of the three possible local magnetic fields due to the bonding electrons. For an individual methyl proton, the probability for a given methylene proton configuration is 1:2:1. In addition, the two spin-up arrangement will be of lowest energy, the one spin up–one

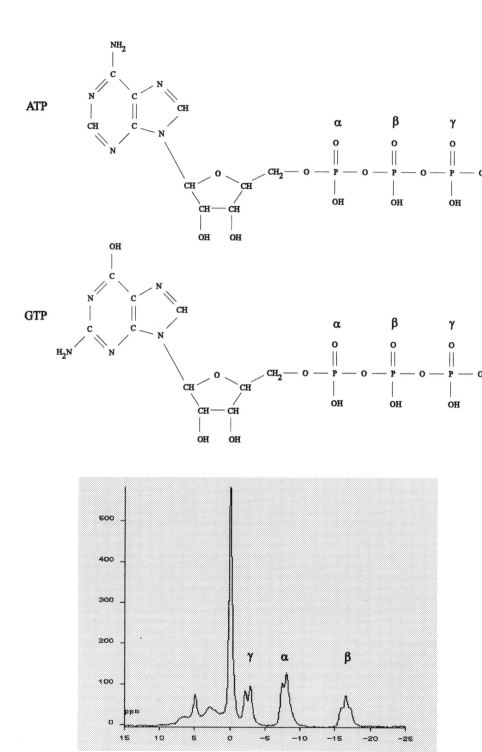

Figure 2.16. Molecular structures for ATP and GTP and ^{31}F spectrum from calf muscle. Phosphorus resonances are equivalent for the two molecules. Chemical shifts are listed in Table 2.3.

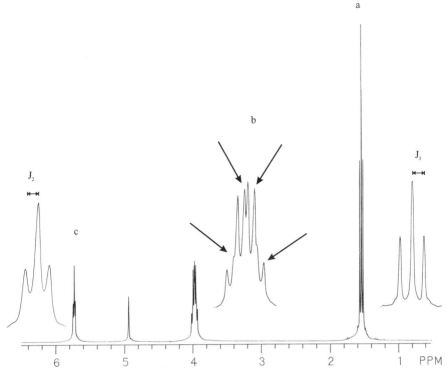

Figure 2.17. Same as Figure 2.15, good spectral resolution. The ethanol resonances are also displayed on an expanded scale to visualize the coupling. Spin-spin coupling between protons on adjacent carbon atoms within ethanol divides the methyl resonance a into a triplet with coupling constant J_1 (due to the methylene protons), the hydroxyl resonance c into a triplet with coupling constant J_2 (due to the methylene protons), and the methylene resonance b into a doublet of quartets (due to the hydroxyl and methyl protons, respectively). One quartet is indicated by the arrows at resonance b, and the other quartet is unmarked. The total relative area of all peaks is 3:2:1.

spin down will be slightly higher in total energy, while the two spin-down arrangement will be the highest energy of the three arrangements. In a collection of protons within a large sample of ethanol, all three local magnetic fields will be present and will generate three resonance absorptions or "peaks," centered at the resonant frequency for the methyl protons. The relative areas of the peaks of the "triplet" resonance will be 1:2:1, due to the greater likelihood for the total spin-zero configuration. The frequency difference between each of the three absorptions is known as the coupling constant J, and is measured in hertz. The methylene protons also couple to the hydroxyl proton. Using the same analysis as before, the single hydroxyl resonance will be divided into three peaks of relative areas 1:2:1 by the two methylene protons [Fig. 2.15(c)]. The coupling constant for this coupling

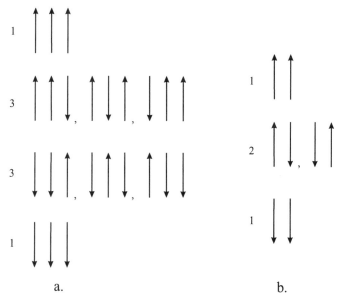

Figure 2.18. (*a*) Probability for spin arrangements of three protons, based on binomial distribution; (*b*) probability for spin arrangements of two protons.

(J_2) will not necessarily be the same value as the one for the methyl-methylene coupling (J_1).

The methyl protons and the hydroxyl proton also couple with the methylene protons. For the three methyl protons, there are four possible arrangements: all three spins up (one possibility, total spin $\frac{3}{2}$), all three spins down (one possibility, total spin $-\frac{3}{2}$), two spins up and one spin down (three possibilities, total spin $\frac{1}{2}$), and two spins down and one spin up (three possibilities, total spin $-\frac{1}{2}$) [Fig. 2.18(*a*)]. The methylene protons will "sense" one of the four configurations, with a probability distribution of 1:3:3:1. The methylene resonance will be divided into four peaks, with relative areas 1:3:3:1. The coupling constant will be the same value as the one for the methyl resonance (J_1 in Fig. 2.17). The coupling of the hydroxyl proton to the methylene protons further divides each peak of the quartet into a doublet, resulting in eight separate resonances. The doublet splitting, J_2, is the same as the splitting of the hydroxyl proton. Instances where more than three identical spins are coupled may be analyzed in a similar fashion using standard probability statistics for a binomial distribution.

There is no requirement that the spins be of the same type for spin–spin coupling to occur. Two important instances of this are ^{1}H–^{13}C and ^{1}H–^{31}P. Many spectral regions in ^{13}C and ^{31}P spectra are complicated by the multiplets generated by spin–spin coupling with the ^{1}H nucleus. The number of multiplets produced by spin–spin coupling is $2n_II + 1$ where n_I is the number of spins of type I involved in the coupling. The value for the coupling constant depends on the number of chemical bonds between the coupled spins. Spins that are directly bonded such as ^{1}H–^{13}C will have larger coupling constants, while coupled spins that are not directly bonded such as ^{1}H–^{1}H will have smaller values for the coupling constant. For most biological systems, the value for the coupling constant is significantly smaller than the chemical shift difference between the two spins that are coupled. This observation allows the two interactions to be treated separately, a so-called "first-order" approximation to the spectral analysis.

Spin coupling is manifest in the MR spectrum as long as the spin under observation senses the coupled nucleus in different spin states. A process known as decoupling allows the MR spectrum to be acquired in the absence of this coupling. Spin decoupling results when the molecular environment of a spin changes rapidly enough between two different states so that the spin cannot distinguish two molecular environments. There are two important instances where this occurs. For some systems, the spins "sense" an average environment due to fast chemical exchange between the two environments (see below). Decoupling also occurs when the coupling spins are excited at their resonant frequency sufficiently for saturation (see below). Because the spin populations are equal in saturation, there is no difference in the magnetic environment and the multiplet resonance of the observed spin will be reduced to a single resonance (Fig. 2.19). Homonuclear decoupling is used when the coupling nucleus is the same type as the observed nucleus (^{1}H decoupling–^{1}H observation). Heteronuclear decoupling is used when the coupling nucleus is a different type from the observed nucleus (^{1}H decoupling–^{13}C observation). Additional hardware is necessary for heteronuclear decoupling. In addition to the reduction in the number of peaks, decoupling may give rise to signal increases or decreases known as the nuclear Overhauser enhancement, described in Section 2.6.

2.6. TIME-DEPENDENT INTERACTIONS

Time-dependent interactions are those in which the spin experiences an interaction that varies with time. The time scale for the variation determines its effect on the MR signal for the spin. Relatively slow processes such as relaxation may be visible in the MR spectrum and can often be analyzed directly. Direct measurement of rapid processes such as fast chemical exchange may not be possible, but often can be inferred from the resulting spectral modifications.

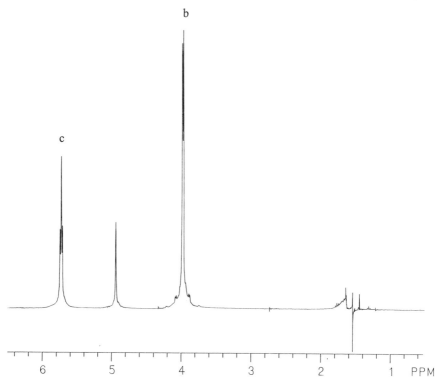

Figure 2.19. Same as Figure 2.17 except that a frequency selective saturation pulse is applied at the frequency for the methyl protons (*a*). The saturated methyl protons are no longer coupled to the methylene protons and the quartet collapses to a doublet due to the hydroxyl coupling.

2.6.1. Relaxation

The previous sections described the process of resonance absorption for the excitation of spins. Equal in importance in MR is the concept of relaxation. Relaxation is the process by which spins release this energy of excitation and return to their original configuration. While an individual spin absorbs energy during excitation, relaxation times are measured for an entire collection of spins and are statistical or average measurements. Relaxation times are measured for a tissue as a bulk sample rather than the individual molecules within the organs. Two measures of relaxation are used, T1 and T2, but both describe energy transfer by the excited spin. Differences in relaxation times between tissues are primarily responsible for contrast in MRI. However, relaxation in MRS complicates the quantitative analysis of spectra to obtain tissue metabolite concentrations. At the same time, analysis of relaxation times can provide useful insights into the molecular environments of spins.

Relaxation time differences can also be used for selective suppression of signals to simplify spectra.

2.6.1.1. T1 Relaxation, Saturation, and the Nuclear Overhauser Effect

The relaxation time T1, also known as the spin-lattice relaxation time or longitudinal relaxation time, is the time required for the z component of **M** to return to 63% of its original value following an excitation pulse. It is the mechanism by which the spins give up their energy to return to their equilibrium orientation. Recall that M_0 is parallel to B_0 at equilibrium and that energy absorption will rotate M_0 into the transverse plane. If a 90° pulse is applied to M_0 as illustrated in Figure 2.10, there will be no longitudinal magnetization following the pulse. As time goes on, a return of the longitudinal magnetization will be observed as the spins release their energy (Fig. 2.20). This return of magnetization follows an exponential growth process, with T1 being the time constant for the growth. After three T1 time periods, **M** will have returned to 95% of its value prior to the excitation pulse, M_0.

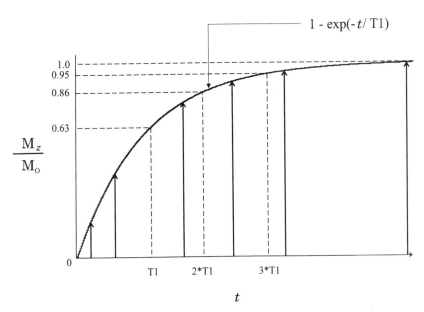

Figure 2.20. The T1 relaxation curve. Following a 90° rf pulse, there is no longitudinal magnetization. Longitudinal magnetization will be regenerated as the protons release their energy through T1 relaxation. Gradually, as more protons release their energy, a larger fraction of M_z is reestablished. Eventually, M_0 will be restored completely. The change of M_z/M_0 with time t follows an exponential growth process. The time constant for this is T1, the spin-lattice relaxation time, and is the time when M_z has returned to 63% of its original value.

The term spin lattice refers to the fact that the excited spin (spin) transfers its energy to its surroundings (lattice) rather than to another spin. The energy no longer contributes to spin excitation.

The key to this energy transfer is the presence of some type of molecular interaction in the vicinity of the excited spin that modulates or varies with an intrinsic frequency ω_L matching the resonant frequency ω_0. The closer ω_0 is to ω_L, the more readily the interaction will absorb the energy and the more frequently this energy transfer will occur, allowing the collection of spins to give up the absorbed energy and return to its equilibrium configuration sooner. In tissues, molecular rotations or tumbling of proteins typically have ω_L approximately 1 MHz. Therefore, these macromolecules will have a better match between ω_0 and ω_L. A more efficient energy transfer will occur and thus the T1 values for protein nuclei will be shorter. For metabolites, which are usually small molecules, the rate of molecular motion is rapid so that there is a poor match between ω_L and ω_0. This means that the T1 relaxation times are relatively long for the molecules studied in clinical spectroscopic studies. Table 2.4 lists approximate T1 relaxation times for some common metabolites.

The nature of the molecular interaction that is modulated is also important to the spin-lattice relaxation process. For spin ½ nuclei, the most common interaction arises from nearby spins. These spins (magnetic dipoles) will affect the local magnetic field around the original spin via a dipole–dipole interaction through space, as long as the spins are near each other (Fig. 2.21).

TABLE 2.4. T1 Relaxation Times for Selected Resonances, 1.5 T

Tissue	T1 (s)
Lactate, methyl group, brain occipital lobe	1.55
NAA, methyl group, brain occipital lobe	1.45
Creatine, N-methyl group, brain occipital lobe	1.55
Choline, N-(Methyl)$_3$ group, brain occipital lobe	1.15
P_i, liver	0.36
α-ATP, liver	0.63
β-ATP, liver	0.65
γ-ATP, liver	0.39
P_i, calf muscle	4.7
PCr	6.5
α-ATP, calf muscle	4.2
β-ATP, calf muscle	4.1
γ-ATP, calf muscle	3.9

[a]The ^{1}H values are from Frahm, et. al[11]. The ^{31}P values are from Buchtal et al.[12]

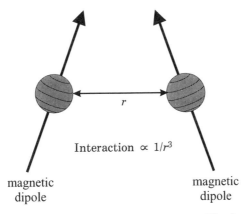

Interaction $\propto 1/r^3$

magnetic   magnetic
dipole   dipole

Figure 2.21. Dipole–dipole interaction between two protons. The force of interaction between the protons is inversely related to the distance between them.

As the dipole pair rotates or vibrates with the motion of the spins, it can absorb the spin excitation energy as part of the "lattice." For spins with $I >$ ½, an additional mechanism is prominent. For these nuclei, the distribution of charge within the nucleus is nonspherical. This produces a quadrupole moment for the nucleus, which is not present for the spherical spin ½ nuclei. For many nuclei (e.g., ^{23}Na), this quadrupole moment is of a significant magnitude. Molecular rotation of this quadrupolar moment will cause these nuclei to undergo T1 relaxation very rapidly.

The consequences of T1 relaxation and the MR signal can be understood through a comparison of the microscopic and macroscopic pictures. Suppose the rf energy is continuously applied at the resonant frequency of the spins so that no relaxation occurs. In the macroscopic picture, the spins in the lower energy level absorb the rf energy and the spins in the upper energy level are stimulated to emit their energy. Since energy is continuously transmitted, the spin populations of the two levels will gradually equalize. When this occurs, no further net absorption of energy is possible, a situation known as saturation. In the macroscopic picture, **M** will rotate continuously but gradually get smaller in magnitude until it disappears as the net population difference approaches zero. Since there is no net magnetization, there will be no coherence in any direction and thus no signal can be produced. This condition is known as saturation. There is a limited amount of energy that spins can absorb before they will become saturated.

In a typical MR experiment, pulsed rf energy is used with a delay time between repeated pulses. This time between pulses allows the excited spins to release the absorbed energy (T1 relaxation). As the spins give up this energy to their surroundings, the population difference between the energy levels is reestablished so that additional absorption can occur during the next rf pulse. In the macroscopic picture, **M** returns toward its initial value **M_0** as more energy is dissipated. Since **M** is the ultimate source of the MR signal, the more energy dissipated, the more signal will be generated following the *next* rf pulse. For practical reasons, the time between successive rf pulses TR is usually insufficient for complete T1 relaxation. The longitudinal magnetization **M** is not completely restored to its equilibrium value, **M_0**. Application of a second rf pulse prior to complete relaxation will rotate **M** into the transverse plane, but with a smaller magnitude than following the first rf pulse. The following experiment describes the situation:

1. A 90° rf pulse is applied. **M** will be rotated into the transverse plane.
2. A time TR elapses, which is insufficient for complete T1 relaxation. The longitudinal magnetization at the end of TR, **M′**, will be less than in Step 1.
3. A second 90° rf pulse is applied. Now **M′** will be rotated into the transverse plane.
4. After a second time TR elapses, **M″** will be produced. It is smaller in magnitude than **M′**, but the difference is less than the difference between **M** and **M′** (Fig. 2.22).

Following a few repetitions, **M** will return to the same magnitude prior to each rf pulse; that is, **M** achieves a steady-state value that depends on the total number of spins in the volume (**M_0**), how efficiently the spins give up their energy (T1 relaxation time), the rate of rf pulse application (time TR), and the amount of energy applied per rf pulse (the pulse amplitude or flip

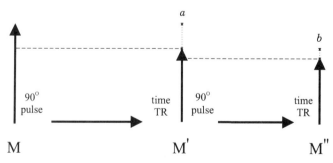

Figure 2.22. Following a 90° rf pulse, longitudinal magnetization is regenerated via T1 relaxation. If the time between successive rf pulses TR is insufficient for complete recovery of **M**, then only **M′** will be present at the time of the next rf pulse (*a*). If time TR elapses again, then only **M″** will be present (*b*). **M″** will be smaller than **M′**, but the difference will be less than the difference between **M** and **M′**.

angle). The detected signal amplitude depends on the value of **M** in this steady state and thus the repetition time TR and T1. In clinical spectroscopic measurements, a steady state of **M** is present because multiple rf pulses are applied and the repetition time TR between each excitation pulse is nearly always insufficient for complete relaxation. To produce this steady state prior to data collection, additional rf pulses are applied immediately prior to the main pulses. These extra rf pulses are known as preparatory pulses or dummy pulses because the generated signals are usually ignored. These preparatory pulses ensure that **M** has the same magnitude prior to every measurement during the scan.

Saturation effects are a particular problem in clinical spectroscopic studies. As mentioned above, most metabolites that are studied using MR spectroscopy have relatively long T1 relaxation times. The short TR times used for spectroscopic studies mean that the spins are saturated to a significant degree and the resulting MR signals are reduced from their "equilibrium" value. This situation makes quantitation of signals difficult since the MR signal is directly proportional to the number of spins only when the collection of spins is at equilibrium. In addition, the degree of saturation differs for each type of spin. Saturation effects must be taken into consideration when attempting to determine concentrations of metabolites from their MR signals (Fig. 2.23).

If the spin under observation is coupled to another spin and this coupling is removed by saturation or "decoupling" techniques (see above), the spin under observation may undergo "cross-relaxation," in which the relaxation rate and the signal amplitude of the observed spin will be modified due to the irradiation of the secondary spin. In biological systems, the most common example of this effect is the nuclear Overhauser enhancement, or NOE.[2] It is observed in ^{1}H decoupled–^{13}C observation or ^{1}H decoupled–^{31}P observation studies and can result in a significant increase in signal amplitude, as much as a threefold increase for the ^{13}C spins attached to the ^{1}H spins.

Consider a molecule containing two unequal spins such as ^{1}H and ^{31}P inside a magnetic field. The Zeeman diagram for the pair of spins will consist of four levels [Fig. 2.24(a)]. Application of a decoupling pulse at the ^{1}H frequency will equalize the spin populations of the ^{1}H spins, producing saturation [Fig. 2.24(b)]. If there is any coupling between the ^{1}H and ^{31}P, either spin–spin (through bonds) or dipole–dipole (through space) in nature, the ^{31}P spin populations will be affected but only between those states where the ^{1}H is unchanged (i.e., only between ^{1}H spin up states or between ^{1}H spin-down states) [Fig. 2.24(c)]. Because the total system is not at equilibrium, the ^{31}P spins will undergo T1 relaxation to reestablish equilibrium. This new equilibrium will have a greater spin polarization (difference in spin up and spin down) of the ^{31}P nuclei than before. Excitation of the ^{31}P spins will

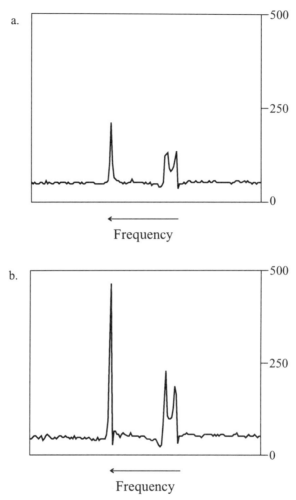

Figure 2.23. Effect of saturation on ¹H MR signals from lactate (doublet) and acetate (singlet). All measurement parameters except TR are the same for the two measurements. Other parameters: Measurement technique: PRESS; TE: 270 ms; Acquisitions: 8; Water suppression used; Data points 1024; Sample width: 1000 Hz. (*a*) TR: 1500 ms and (*b*) TR: 5000 ms.

produce a 50% greater signal than would be expected simply from a decoupled spectrum. This signal increase is the NOE. The extent of enhancement depends on the particular nuclei involved, the relative difference in γ values between the nuclei, and the extent of dipolar relaxation.

Three methods are commonly used to measure T1 relaxation times. One method, progressive saturation, uses a 90° rf pulse followed by signal detec-

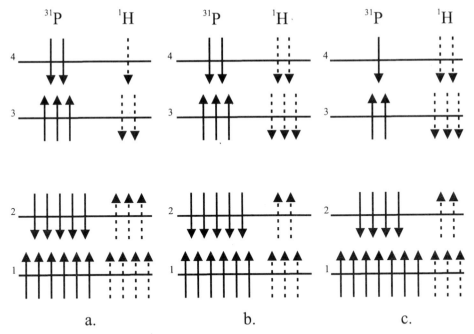

a.   b.   c.

Figure 2.24. Nuclear Overhauser enhancement. (a) Unperturbed equilibrium arrangement of coupled ^{31}P spins (solid arrows) and ^{1}H spins (dashed arrows) inside a magnetic field (not shown); (b) ^{1}H spins are saturated, equalizing spin populations between levels 1 and 3 and between levels 2 and 4: (c) new equilibrium arrangement is established with increased spin polarization (greater difference in the number of spins) for the ^{31}P spins (levels 1 and 2).

tion. The time between repetitions of the pulse-detection process is varied [Fig. 2.25(a)]. The growth of signal amplitude is determined by the T1 relaxation time. This process is simple and relatively rapid in execution, but requires accurate 90° pulses and that only longitudinal magnetization is present at the time of the excitation pulse. The other two methods consist of two rf pulses per cycle, one that modifies the longitudinal magnetization for the spins, followed by a 90° rf pulse to rotate the modified magnetization into the transverse plane and produce a signal. The time between the pulses is varied to provide different time points for calculation of the T1 relaxation time. The first approach, saturation recovery, applies sufficient rf energy during the first pulse to saturate the spins [Fig. 2.25(b)]. The growth of longitudinal magnetization and thus the signal amplitude following the second pulse is determined by T1. While simple in concept, this approach requires complete saturation of the spins, which is often not realized in actual circumstances. The preferred approach for measuring T1 relaxation times is the Inversion recovery method, where the initial pulse is a 180°

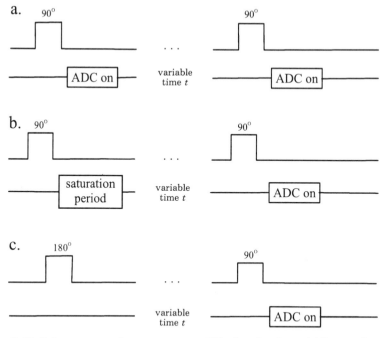

Figure 2.25. Pulse sequences for measurement of T1 relaxation times. (*a*) Progressive satura-tion. Signal amplitude is measured at different times *t* between measurements. (*b*) Saturation recovery. Signal amplitude is measured at different times *t* following spin saturation. (*c*) Inversion recovery. Signal amplitude is measured at different times *t* between inversion pulse and detection pulse.

inversion pulse [Fig. 2.25(*c*)]. While higher amounts of power and very long TR are required, incomplete inversions due to rf pulse imperfections do not cause distortions of the signal amplitude.

2.6.1.2. T2 Relaxation, T2* Relaxation, and Spin Echoes

The relaxation time T2, also known as the spin–spin relaxation time or transverse relaxation time, is the time required for the transverse component of $\mathbf{M}$ to decay to 37% of its initial value via irreversible processes. As described earlier, $\mathbf{M_0}$ is oriented only along the z ($\mathbf{B_0}$) axis at equilibrium and no portion of $\mathbf{M_0}$ is in the xy plane. The coherence or uniformity of the spins is entirely longitudinal with no transverse component. Absorption of energy from a 90° rf pulse as in Figure 2.10 causes $\mathbf{M_0}$ to rotate entirely into the xy plane, so that the coherence is in the transverse plane at the end of the pulse. As time elapses, this coherence disappears while at the same time the spins release their energy and reorient themselves along $\mathbf{B_0}$. This disappearing coherence produces the FID described above. As this coherence disappears, the value of $\mathbf{M}$ in the xy plane ($\mathbf{M}_{xy}$) decreases toward zero. The process by which this transverse magnetization is lost is either T2 or T2* relaxation.

A comparison of microscopic and macroscopic pictures provides additional insight into this process. At the end of the 90° rf pulse, when the spins have absorbed energy and $\mathbf{M}$ is oriented in the transverse plane, each spin precesses at the same frequency ω_0 and is synchronized at the same point or phase of its precessional cycle. Since a nearby spin of the same type will have the same molecular environment and the same ω_0, it will readily absorb the energy that is being released. Spin–spin relaxation refers to this energy transfer from an excited spin to another nearby spin. The absorbed energy remains as spin excitation rather than being transferred to the surroundings as in T1 relaxation. This energy transfer between spins can occur many times as long as the spins remain at the same ω_0 and are in close proximity to each other. Intermolecular and intramolecular interactions such as vibrations or rotations will cause the local magnetic field to fluctuate or modulate around each spin causing ω_0 to vary. This fluctuation will produce a gradual, irreversible loss of phase coherence to the spins as they exchange the energy and reduce the magnitude of the transverse magnetization (Fig. 2.26). T2 is the time when the transverse magnetization is 37% of its value immediately after the 90° pulse when this irreversible process is the only cause for the loss of coherence. As more time elapses, this transverse coherence completely disappears only to reform in the longitudinal direction as T1 relaxation occurs. This intrinsic dephasing time T2 will always be less than or equal to T1.

There are several potential causes for a loss of transverse coherence to $\mathbf{M}$. One is the modulation of the local magnetic field of the spin due to molecular vibrations or rotations. This is responsible for spin–spin relaxation or the true T2. In many situations, T2 relaxation can be used as a measure of the restriction of molecular motion of the spin. Short T2 relaxation times may indicate restricted motion while long T2 times are consistent with more freedom or randomness to the spin movement. For spins that are bound to macromolecules such as protons within the protein backbone or metabolites that are strongly adsorbed, the T2 relaxation times are too short to enable observation of the signal. The "visible" signals from metabolites are from those molecules with relatively unrestricted motion and long T2 times. For nuclei with significant quadrupole moments, such as ^{23}Na, the modulation of the nuclear quadrupole moment shortens T2 significantly. Another cause for the loss of transverse coherence arises from the fact that a spin never experiences a magnetic field that is 100% uniform or homogeneous. As the spin precesses, it experiences a fluctuating local magnetic field, causing a change in ω_0 and a loss in transverse phase coherence. This nonuniformity in $\mathbf{B}_0$ comes from three sources:

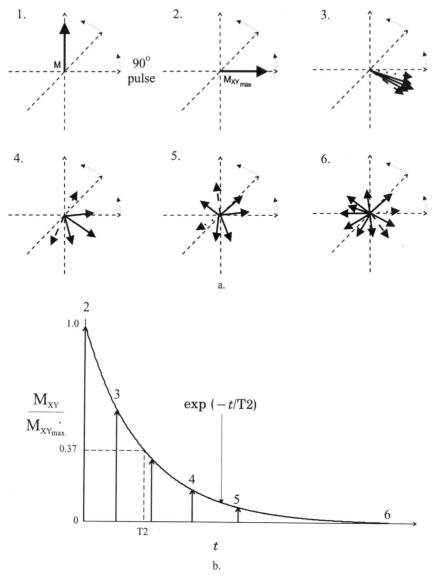

Figure 2.26. T2 Relaxation. (*a*) A rotating frame slower than ω_0 is assumed for this figure. Prior to the excitation pulse, the net magnetization $\mathbf{M_0}$ is parallel to $\mathbf{B_0}$ (not shown) (1). Immediately following a 90° rf pulse, the protons are initially precessing in phase in the transverse plane (2). Due to inter- and intramolecular interactions, the protons will begin to precess at different frequencies (dashed arrow = faster; dotted arrow = slower) and will become asynchronous with each other (3). As more time elapses (4, 5), the transverse coherence becomes smaller until there is complete randomness of the transverse components and no coherence (6). (*b*) Plot of relative $\mathbf{M_{XY}}$ component. The numbers correspond to the expected $\mathbf{M_{XY}}$ component from (*a*). The change in $\mathbf{M_{XY}}/\mathbf{M_{XY_{max}}}$ with time follows an exponential decay process. The time constant for this process is the spin–spin relaxation time T2 and is the time when $\mathbf{M_{XY}}$ has decayed to 37% of its original value.

1. Main field inhomogeneity. There will always be some degree of non-uniformity to $\mathbf{B_0}$ due to imperfections in magnet manufacturing, composition of nearby building walls, or other external sources of metal.

2. Sample induced inhomogeneity. Differences in the magnetic susceptibility or degree of magnetization of adjacent tissues (e.g., bone or air) will distort the local magnetic field near the interface between the different tissues.

3. Localization gradients. The technique used for spatial localization generates a magnetic field inhomogeneity that induces spin dephasing. Proper design of the pulse sequence will eliminate the localization gradients as a source of dephasing.

These other sources will contribute to the total transverse relaxation time, T2*:

$$1/T2* = 1/T2 + 1/T2_M + 1/T2_{MS} \tag{2.8}$$

where $T2_M$ is the dephasing time due to the main field inhomogeneity and $T2_{MS}$ is the dephasing time due to the magnetic susceptibility differences. For spectroscopic studies away from tissue–air or tissue–bone interfaces, $T2_{MS}$ does not contribute significantly. For most studies of spin ½ nuclei or for nuclei with small quadrupole moments (e.g., ^{7}Li), $T2_M$ predominates so that the decay of the transverse magnetization following a 90° rf pulse, the FID, is determined by T2* rather than just T2. For nuclei with significant quadrupole moments (e.g., ^{23}Na), T2 is usually very short and is the predominant term in Eq. 2.8.

Some of these sources of spin dephasing can be reversed by the application of a 180° rf pulse.[3] This is illustrated by the following sequence of events (Fig. 2.27):

1. A 90° rf pulse, which rotates $\mathbf{M}$ into the transverse plane.
2. A short delay of time t, which allows $\mathbf{M}$ to dephase.
3. A 180° rf pulse, which rotates $\mathbf{M}$ about the axis of the pulse.
4. Another delay of time t, which allows $\mathbf{M}$ to rephase.

During the time t, spin dephasing will occur through T2* relaxation processes. Application of the 180° rf pulse causes the spins to effectively reverse their phases relative to the resonant frequency. The rates and directions of precession for the spins do not change following the 180° rf pulse, only their relative phases. If time t elapses again, then the spins will regain their transverse coherence. This reformation of phase coherence induces a voltage in the receiver coil, which is known as a spin echo. Sources of dephasing that do not change during the two time periods such as the main field inhomo-

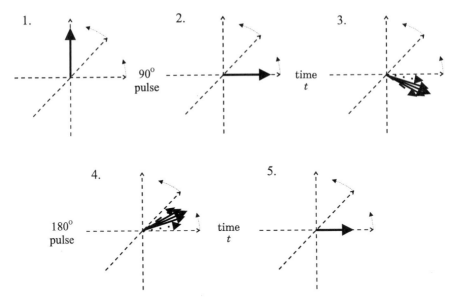

Figure 2.27. A rotation frame slower than ω_0 is assumed for this figure. Application of a 90° rf pulse will rotate $\mathbf{M_0}$ into the transverse plane (2). Due to T2* relaxation processes, the protons will become asynchronous with each other during time t (3). Application of a 180° rf pulse will cause the protons to reverse their phase relative to the transmitter phase. The protons that precessed most rapidly will be farthest behind (dashed arrow) while the slowest protons will be in front (dotted arrow). Allowing time t to elapse again will allow the protons to regain their phase coherence and form a spin echo (5). The loss in magnitude of the echo relative to the original coherence (2) is due to irreversible processes (i.e., true spin–spin or T2 relaxation).

geneity and the magnetic susceptibility differences, will exactly cancel since the spin will experience exactly the same interactions. This means that the contributions to T2* relaxation from these static sources will disappear. Only the irreversible spin-spin relaxation is unaffected by the 180° rf pulse so that, for a spin echo, the loss of phase coherence and signal amplitude is due only to true T2 relaxation.

The above description is slightly complicated by the addition of chemical shift and spin–spin coupling interactions. Spin dephasing due to chemical shift interactions is rephased by the 180° rf pulse in the same fashion as described above for main field inhomogeneity and magnetic susceptibility differences. For spins that are coupled, the 180° rf pulse will not reverse dephasing effects. Coupled spins always maintain their frequency difference regardless of any rf pulse or the static magnetic field (recall that spin–spin coupling is independent of $\mathbf{B_0}$). The result is a phase modulation to the signal that depends on the time t between the rf pulses. The modulation frequency

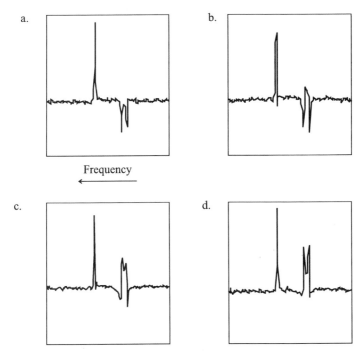

Figure 2.28. Effect of TE on PRESS spectra from sample of acetate and lactate. Other measurement parameters: TR 1500 ms; Number of acquisition 8; Vector size 1024 data points. (*a*) TE 135 ms; (*b*) TE 170 ms; (*c*) TE 202 ms; (*d*) TE 270 ms. Note acetate resonance (singlet) phase is unchanged relative to lactate resonance (doublet).

is proportional to $1/J$. Proper choice of the rf pulse timing is necessary to ensure that the spins are in phase at the time of measurement. The difficulty is that no one choice of pulse spacing will ensure that all spins under observation will be in phase at the same time since there is no single coupling constant for all spins.

The particular choice of echo time, TE, the time between the excitation pulse and echo maximum, complicates the spectrum only if the spins are coupled. Figure 2.28 illustrates this problem. The methyl and methine protons of lactate are coupled with a coupling constant of approximately 7 Hz. This results in a phase modulation of the lactate proton signal with a period of 270 ms.[d] At TE = 135 ms [Fig. 2.28(*a*)], the signal will be negative relative to a noncoupled acetate resonance. At other TEs, the lactate signal will be modulated while the acetate signal will only be reduced in amplitude.

[d]Historically, a TE of 270 ms was used, based on a coupling constant of 7 Hz. More accurate measurements of the coupling constant yield a value of 6.93 Hz ($1/J$ = 144 ms).[4] This is only a problem when lactate is explicitly studied.

Following the echo formation, the spins continue to precess and will dephase a second time since the sources of dephasing continue to affect them. Application of another rf pulse again reverses the spin phases and generates another spin echo. In fact, several echoes are generated following this third rf pulse. The number of echoes, their amplitude, and their timing relative to the first pulse depend on the relative spacing and the amplitudes of the individual rf pulses (Fig. 2.29). Such multiple rf pulse schemes are also used in standard MR imaging. In Figure 2.29, the echoes labeled 1 and 2 are typically used to produce proton density weighted and T2-weighted images, respectively. These echoes are maximized when rf pulse 1 is a 90° excitation pulse and pulses 2 and 3 are 180° refocusing pulses. Echo 2 is typically used in the PRESS localization scheme for in vivo spectroscopy (see Section 4.4.2.3). Echo 3 is known as a stimulated echo. It is the result of all three rf pulses acting together. Some of the transverse magnetization generated by pulse 1 will be returned to the longitudinal direction following pulse 2. Pulse 3 rotates this longitudinal magnetization into the transverse direction and produces an echo, the stimulated echo. Its amplitude is a maximum when all three rf pulses are 90° excitation pulses. Stimulated echoes are used in the STEAM localization technique for in vivo spectroscopy (see Section 4.4.2.2). Stimulated echoes have the advantage over spin echoes in that they are not modulated by J coupling, making their amplitude less sensitive to the choice of TE. The spectra can be phase corrected to give all positive amplitude signals. Stimulated echoes may be produced with less transmitted rf energy than a spin echo. However, the maximum amplitude for a stimulated echo will be one-half of that for a spin echo acquired at the same TE. This intrinsic signal reduction may render stimulated echo-based techniques unsuitable for some spectroscopic applications.

The T2 relaxation times may be measured by performing a series of spin echo measurements, each with a different TE time. The simplest technique uses two pulses, a 90–180° pair, and simply increases the time between the pulses [Fig. 2.30(a)]. This produces an echo known as a Hahn echo. While Hahn echoes are used in MR imaging to minimize the rf power deposition and gradient requirements, they are sensitive to diffusion effects that cause enhanced spin dephasing and signal loss as well as to pulse imperfections (inaccurate or incomplete rotations). More accurate approaches use multiple 180° refocusing pulses with a constant pulse spacing. The number of refocusing pulses between excitation and detection is varied, thereby changing the TE for the detected signal. Short interpulse spacing minimizes diffusion contributions to the signal decay. The Carr–Purcell approach uses the same phase for all pulses [Fig. 2.30(b)]. The most accurate technique, the Carr–Purcell–Meiboom–Gill modification, is the same as the Carr–Purcell method except that the refocusing pulses are phase-shifted 90° from the excitation

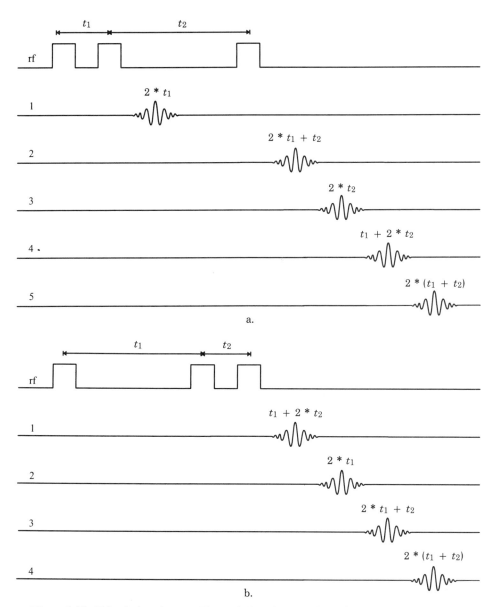

Figure 2.29. Echo timing plots. (a) Time t_1 is less that t_2, typical of a standard short TE/long TE spin echo pulse sequence or PRESS localization sequence. Five echoes are formed. Two are used in routine imaging (1 and 3). The stimulated echo occurs at time $2t_1 + t_2$ (echo 2). (b) Time t_1 is greater than t_2, typical of a spin echo pulse sequence with a short TE and a spatial presaturation pulse or a STEAM localization sequence. Four echoes are formed. Two are used in routine imaging (1 and 2). The stimulated echo occurs at time $2t_1 + t_2$ (echo 3).

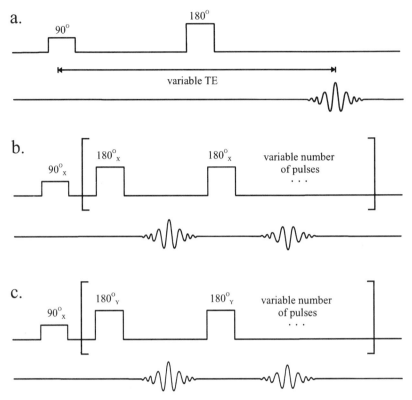

Figure 2.30. Pulse sequences for measurement of T2 relaxation times. (*a*) Hahn echo sequence: T2 variation is produced by varying the TE. (*b*) Carr–Purcell sequence: T2 variation is produced by varying the number of 180° pulses. Phase of signal alternates every other pulse. (*c*) Carr–Purcell–Meiboom–Gill sequence: Same as (*b*) except the 180° pulses are phase shifted relative to the 90° pulse. Phase of signal is constant.

pulse [Fig. 2.30(*c*)]. Errors from pulse imperfections are eliminated following every second refocusing pulse by this approach, enabling a more accurate estimate of T2 using only the amplitudes of the even echoes.

2.6.2. Chemical Exchange

Another time-dependent process that can affect the MR spectrum is known as chemical exchange.[5] Exchange processes are a group of phenomena in which the spins of interest change molecular environments or collide and change spin states. There must be a difference either in the two particles, the two molecular environments, or both in order to observe the process. The most important example of this process in biological MR spectroscopy is chemical exchange, where a spin alternates between two different environments. Exchange differs from T2 relaxation in that T2 relaxation involves only energy transfer between the two states while exchange requires an environmental difference between the two states. Exchange processes are characterized by their rate relative to the MR measurement process. Slow exchange rates are those in which the interchange of environments is slow enough to enable resolution of signals from each environment. Fast exchange rates occur rapidly enough that only one signal is observed.

Exchange processes alter the MR spectra by affecting both the resonant frequencies and line widths of the species undergoing exchange relative to the nonexchanging situation. The nature of the changes depends on the exchange rate relative to the frequency separation of the resonances. Two spins undergoing slow exchange have two resonant frequencies that approach each other and line widths that broaden as the exchange rate increases. The exact position and width of each signal depends on the relative concentration and the exchange rate. Two spins undergoing fast exchange produce one signal at a frequency that is the mean of the individual frequencies weighted by the concentrations of the two spins. The line width narrows as the exchange rate increases.

A simple example of this exchange is illustrated in the ethanol solution used in Figure 2.17 following the addition of a small amount of acid. The presence of the acid allows the water and hydroxyl protons to exchange places (Figure 2.31). This exchange happens rapidly enough that neither proton is exclusively on the water or the ethanol molecule during the measurement, but each spends part of the time on both molecules. The result is that both resonances are broadened and move together (Figure 2.32). In

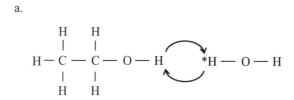

Figure 2.31. Water–ethanol exchange diagram. The alcohol proton and a water proton will interchange molecules so that each proton will sense both molecular environments.

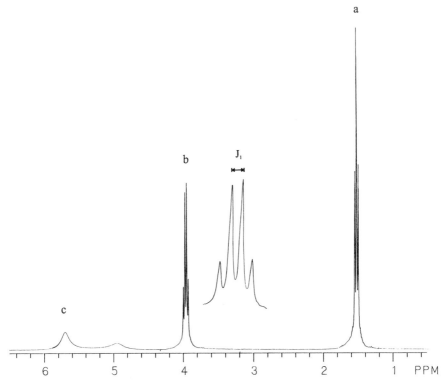

Figure 2.32. Same as Figure 2.17, except for the addition of a few drops of acid. The water and hydroxyl protons (unmarked and (c), respectively) undergo slow exchange. Their resonances are broadened and their separation is reduced. In addition, the spin coupling between the hydroxyl proton and the methylene protons (b) is lost and the methylene protons are represented by a quartet due to coupling with the methyl protons (label a, coupling constant J_1).

addition, the spin coupling between the hydroxyl and the methylene protons is lost since the methylene protons do not "sense" only the hydroxyl proton, but an "average" proton. The result is that the methylene resonance is reduced to a quartet due to the coupling with the methyl protons, and the hydroxyl proton is a single peak.

An important example of fast exchange is the chemical shift variation of the inorganic phosphate (Pi) resonance observed in ^{31}P spectra induced by pH changes within the tissue. At physiological pH ranges, Pi is present at equilibrium in the mono- and diprotonated forms:

$$HPO_4^{2-} + H^+ \rightleftarrows H_2PO_4^-$$

The two phosphate ions have different chemical shifts, due to the presence or absence of the extra proton. The equilibrium transfer of the H^+ ion occurs rapidly enough that an average phosphate environment is present during the measurement. The ^{31}P MR signal for Pi is an average signal from both species. This average is weighted by the relative concentration of each species. As the pH of the tissue changes, the concentration of one of the phosphate ions changes at the expense of the other one. This change in concentration will shift the frequency of the ^{31}P phosphate signal. For typical changes of pH in tissue, the change in chemical shift is approximately 2 ppm.[6] This method is commonly used for monitoring intracellular pH changes following exercise in patients with some types of inborn metabolism abnormalities.

REFERENCES

1. C. P. Slichter, *Principles of Magnetic Resonance*, 3rd ed., Springer-Verlag, New York, 1990.

2. J. H. Noggle and R. E. Schirmer, *The Nuclear Overhauser Effect: Chemical Applications*, Academic, New York, 1971.

3. E. L. Hahn, *Spin Echoes*, *Phys. Rev.* **80**, 589–594, 1950.

4. P. B. Kingsley, *Are You Still Using TE = 136 and 272 for Lactate? Scalar Coupling and ZQC Relaxation in STEAM*, Abstract, 12th Annual Meeting, Society of Magnetic Resonance in Medicine, New York, 1993.

5. J. I. Kaplan and G. Frankel, *NMR of Chemically Exchanging Systems*, Academic, New York, 1980.

6. R. B. Moon and J. H. Richards, Determination of intracellular pH by ^{31}P magnetic resonance. *J. Biol. Chem.* **248**, 7276–7278, 1973.

7. I. M. Mills, *Quantities, Units and Symbols in Physical Chemistry*, International Union of Pure and Applied Chemistry, Physical Chemistry Division, Blackwell, Oxford, UK, 1989.

8. D. Shaw, *Fourier Transform NMR Spectroscopy*, 2nd ed., Elsevier, New York, 1984.

9. J. Frahm, H. Bruhn, M. L. Gyngell, K. D. Merboldt, W. Haenicke, and R. Sauter, Localized high-resolution proton NMR spectroscopy using stimulated echoes: Initial applications to human brain *in vivo*. *Magn. Reson. Med.* **9**, 79–93, 1989.

10. C. T. Burt, S. M. Cohen, and M. Barany, *Annu. Rev. Biophys. Bioeng.* **8**, 1–25, 1979.

11. J. Frahm, H. Bruhn, M. L. Gyngell, K. D. Merboldt, W. Hänicke, and R. Sauter, Localized proton NMR spectroscopy in different regions of the human brain *in vivo*. Relaxation times and concentrations of cerebral metabolites, *Magn. Reson. Med.* **11**, 47–63, 1989.

12. S. D. Buchthal, W. J. Thoma, J. S. Taylor, S. J. Nelson, and T. R. Brown, In vivo T1 values of phosporus metabolites in human liver and muscle determined at 1.5 T by chemical shift imaging, *NMR Biomed.*, **2**, 298–304, 1989.

CHAPTER 3

Instrumentation

The instrumentation used to perform a clinical MR spectroscopic evaluation has evolved over the years and has benefited from the hardware improvements developed to address clinical imaging needs. At the present time, most spectroscopic software provided by the manufacturers of whole-body imaging systems allows both preprogrammed acquisitions and analyses as well as user-controlled operations. Specific questions regarding software features should be addressed to the particular manufacturer. While many of the hardware requirements for clinical spectroscopic systems are the same as for clinical MR imaging systems, some specifications are most stringent and may preclude some examinations from being performed on certain systems. Because most in vivo spectroscopic studies are performed on clinical MR imaging systems, the following description focuses on the fundamental features of a clinical scanner with special emphasis on those aspects pertinent for spectroscopic examinations. For any clinical MR system, the major hardware components are a magnet, a gradient system, an rf system, a computer/array processor, and a data acquisition system (Fig. 3.1).

3.1. MAGNET SYSTEM

The magnet is the primary component of an MR scanner and the most important component for in vivo spectroscopic studies. Magnets are available in a variety of field strengths, shapes, and materials. Magnet field strengths are measured in units of tesla (T) or gauss (G) (1 T = 10,000 G) and clinical MR systems are usually categorized as low, medium, or high-field systems. Low-field magnets have field strengths less than 0.5. T. Medium-field systems have fields between 0.5 and 1.0 T, while high-field systems have fields of 1.0 T or greater. While imaging systems are available using low-field magnets, the small net magnetization generated by these systems makes them impractical for spectroscopic studies: the reduced signal necessitates a tremendous number of acquisitions producing long measurement times. While clinical spectroscopic studies have been performed at 1.0 T or

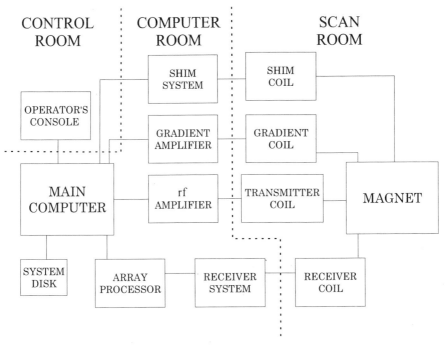

Figure 3.1. Block diagram of an MR scanner.

lower field strengths, most in vivo studies are performed at 1.5 T or higher providing improved spectral separation of the resonances as well as increased S/N. Many whole-body research systems have magnetic fields up to 4 T. High-field magnets are solenoidal superconducting magnets using as the magnet wire niobium–titanium alloy wire immersed in liquid helium. This alloy has minimal resistance to the flow of electrical current below a temperature of 20 K. The cryostat of the superconducting magnet, which contains the liquid helium, may be a double dewar design with a liquid nitrogen container surrounding the helium container, or a helium-only design with a refrigeration system to minimize the helium boiloff.

The primary characteristic of importance in magnet quality is the homogeneity or uniformity of the magnetic field. Field homogeneity is usually measured in ppm relative to the main field over a certain distance. Equation (2.7) can be used to calculate the ppm scale, replacing the frequencies with the measured magnetic field values. High homogeneity means the magnetic field changes very little over the specified region or volume. Identical spins in this region will resonate at the same frequency and thus produce the maximum possible signal. Great effort is taken during magnet manufacturing and installation to ensure the best field homogeneity possible. Most MRI systems use a set of coils known as a shim coil set to compensate for field distortions due to manufacturing imperfections or the environment at individual installations (e.g., nearby steel posts or asymmetrical metal arrangements). The design of the shim coil may be passive in that it holds pieces of metal (shim plates) that correct the field distortions, or active in that there are loops of wire through which current passes to produce supplementary magnetic fields. In some systems, both types of shim correction may be used with the passive shim providing coarse field corrections and the active shim providing fine adjustments.

Field homogeneity and the ability to improve it are important factors to consider when evaluating an MR system for clinical spectroscopy. While S/N is adversely affected by poor homogeneity, spectral resolution is particularly sensitive to field homogeneity. As mentioned in Chapter 2, the line width of a resonance peak is inversely proportional to T2*, so that good field homogeneity allows resolution of resonances with frequencies that are close together. For clinical spectroscopic studies, shimming or optimization of the field homogeneity must be done to the volume of interest. There are two reasons for this. Most magnetic fields are optimized by the manufacturer using uniform phantoms of spherical or cylindrical shape and the (nonspherical) patient distorts the magnetic field. In addition, within the patient are areas of different magnetic susceptibility. Volumes defined in one region of the patient will not necessarily be in the same total magnetic field as those defined in another region.

There are a variety of approaches used for magnet shimming. One method uses a small water sample to "map" the magnetic field by measuring the ^1H resonant frequency at a variety of locations. This technique, known as field mapping, is typically performed by the manufacturer during system installation. A desired field shape (spherical) is chosen and the water frequency is measured from the sample as it is positioned at various points on this sphere. A multilinear regression of the frequencies as a function of the sample positions is performed to determine the deviations from the ideal spherical field shape and the corrections necessary to achieve it. The mathematical functions used in the regression are known as spherical harmonics or Legendre polynomials. The number of functions used in the regression analysis typically depends on the number of measurements made and the degree or order of field correction to be made. Table 3.1 lists some of the spherical harmonic functions (Fig. 3.2). Field corrections are made in a variety of ways. Coarse adjustments may be made using superconducting shim coils, pieces of metal known as shim plates, or using small magnets. For fine adjustments, many MR systems have shim cabinets and coils enabling second or higher order corrections to be made. First-order corrections are usually performed with gradient offset currents (see below).

For shimming the magnetic field to the individual patient or volume, the large ^1H MR signal from the tissue water is used. Shimming on the water

TABLE 3.1. Spherical Harmonic Functions and Cartesian Equivalents

Order of Harmonic	Spherical Harmonic	Spherical Notation[a]	Cartesian Notation
First order	A (1,0)	$r \cos \theta$	z
	A (1,1)	$r \sin \theta \sin \phi$	y
	B (1,1)	$r \sin \theta \cos \phi$	x
Second order	A (2,0)	$(3r^2 \cos^2 \theta - 1)/2$	$(3z^2 - 1)/2$
	A (2,1)	$3r^2 \cos \theta \sin \theta \sin \phi$	$3zy$
	B (2,1)	$3r^2 \cos \theta \sin \theta \cos \phi$	$3zx$
	A (2,2)	$3r^2 \sin^2 \theta \sin 2\phi$	$3(x^2 - y^2)$
	B (2,2)	$3r^2 \sin^2 \theta \cos 2\phi$	$6xy$
Third order	A (3,0)	$(5r^2 \cos^3 \theta - 3r \cos \theta)/2$	$(5z^3 - 3z)/2$
	A (3,1)	$3r^3 \sin \phi \sin \theta (5 \cos^2 \theta - 1)/2$	$3y(5z^2 - 1)/2$
	B (3,1)	$3r^3 \cos \phi \sin \theta (5 \cos^2 \theta - 1)/2$	$3x(5z^2 - 1)/2$
	A (3,2)	$15r^3 \sin 2\phi \sin^2 \theta \cos \theta$	$30xyz$
	B (3,2)	$15r^3 \cos 2\phi \sin^2 \theta \cos \theta$	$15z(x^2 - y^2)$
	A (3,3)	$15r^3 \sin 3\phi \sin^3 \theta$	$15y(3x^2 - y^2)$
	B (3,3)	$15r^3 \cos 3\phi \sin^3 \theta$	$15x(x^2 - 3y^2)$
Fourth order	A (4,0)	$(35r^4 \cos^4 \theta - 30r^2 \cos^2 \theta + 3)/8$	$(35z^4 - 30z^2 + 3)/8$
	A (4,1)	$5r^4 \sin \theta \sin \phi (7 \cos^3 \theta - 3 \cos \theta)/2$	$5y(7z^3 - 3z)/2$
	B (4,1)	$5r^4 \sin \theta \cos \phi (7 \cos^3 \theta - 3 \cos \theta)/2$	$5x(7z^3 - 3z)/2$
	A (4,2)	$15r^4 \sin^2 \theta \sin 2\phi (7 \cos^2 \theta - 1)/2$	$15xy(7z^2 - 1)$
	B (4,2)	$15r^4 \sin^2 \theta \cos 2\phi (7 \cos^2 \theta - 1)/2$	$15(x^2 - y^2)(7z^2 - 1)/2$
	A (4,3)	$105r^4 \sin^3 \theta \sin 3\phi \cos \theta$	$105yz(3x^2 - y^2)$
	B (4,3)	$105r^4 \sin^3 \theta \cos 3\phi \cos \theta$	$105xz(x^2 - 3y^2)$
	A (4,4)	$105r^4 \sin^4 \theta \sin 4\phi$	$420xy(x^2 - y^2)$
	B (4,4)	$105r^4 \sin^4 \theta \cos 4\phi$	$105(x^4 - 6x^2y^2 + y^4)$

[a]The parameter r is the distance away from the magnet isocenter, θ is the angle away from parallel to **B**, the parameter ϕ is the azmuthal angle rotated away from the horizontal direction counterclockwise. [After F. Roméo and D. Hoult, *Magn. Reson. Med., 1*, 44, 1984.]

resonance is performed even for heteronuclear studies. The rationale is that the water signal is easily visible and optimizing the magnetic field of the water protons will improve the signal for the metabolites as well. Current state-of-the-art MR systems have automated shimming procedures using software provided by the manufacturer. There are two methods that are typically used for optimizing the homogeneity. The simplest technique maximizes the FID or echo signal. As mentioned in Chapter 2, the decay of the FID for spin ½ nuclei is dominated by $T2_M$, the main field inhomogeneity. Improving the field homogeneity reduces the $T2_M$ contributions to the FID decay, thereby increasing the signal amplitude and spectral resolution. The second approach uses a magnetic field gradient (see below) to deliberately distort the field homogeneity. The phase variation of the MR signal in the presence of the gradient is proportional to the field inhomogeneity. Optimal

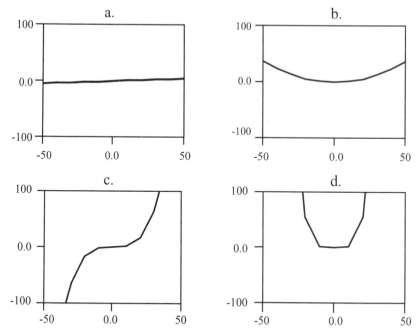

Figure 3.2. Field shapes for spherical harmonics. All curves are plotted with equivalent scales. Horizontal axis: distance from isocenter; vertical axis: effective field variation. (*a*) $A(1,0)$; (*b*) $A(2,0)$; (*c*) $A(3,0)$; $A(4,0)$.

homogeneity is obtained when the phase variation produced by the gradient is constant. Regardless of the method used, shimming to the volume of interest is a vital step in obtaining usable spectroscopic results. For example, field inhomogeneity of less than 0.1 ppm for the volume under observation is necessary to resolve the ^1H resonances of choline and creatine.

There are two situations where shimming may be insufficient for producing adequate localized field homogeneity for clinical spectroscopy. In areas of the body where magnetic susceptibility differences are significant, such as near the frontal sinuses or in the neck, the air–tissue interface produces large asymmetric field distortions that cannot be eliminated. Metal-containing areas such as dental braces or shunt reservoirs cause severe distortions, affecting areas of tissue far away from the metal. Another situation requiring caution is the use of contrast agents based on modification of the T1 or T2 relaxation times. The relaxing agent, usually Gd^{3+}, Mn^{2+}, or Fe^{3+} ions in chelated form, will also affect the local magnetic susceptibility in tissues where it accumulates, decreasing T2* and broadening the spectral peaks. Care should be exercised when performing MR spectroscopic studies following the administration of such agents (see Chapter 6).

Certain precautions should be exercised around all magnets, regardless of field strength. The examination room where the magnet is located should have restricted access. Any metal objects near the magnet should be non-magnetic due to potential interaction with the magnetic field. Strong attraction of the metal object by the magnet can occur, sometimes in an uncontrollable fashion. Stethoscopes or pens may be pulled into the magnet causing possible injury. The force of attraction increases with the mass of the object, necessitating greater distances away from the magnet for large objects. In addition, the magnet may interact with the operation of nearby electrical equipment such as patient monitoring units. Magnetic shielding may be required for proper function. Patients with surgical implants or metal fragments in their bodies as a result of trauma or occupation (e.g., sheet metal workers) should be scanned only if there is no risk to the patient should the implant or fragment move during the procedure. Patients with electronic implants such as pacemakers or ferromagnetic intracranial aneurysm clips should not be scanned under any circumstances due to the risk of patient injury.

It is also important to realize that the magnetic field of all magnets extends in all directions away from the center of the field. The amount of fringe magnetic field, the portion outside the magnet housing, is a very important consideration in siting an MRI system. The fringe field is greatest near the magnet parallel to the main field and decreases with increasing distance away from the magnet. The fringe field is also larger for higher field magnets. Low-field magnets have a very small fringe field, making it easier to use standard patient monitoring equipment. High-field systems are often installed with magnetic shielding of some type to reduce the fringe field. It may surround the magnet (passive shielding), be generated by a second set of superconducting magnet windings opposing the main field (active shielding), or be built into the wall (room shielding). Shielding must be in place at the time that shimming is performed. Two distances are of concern regarding the fringe field. The 0.5 mT (5 G) distance is considered the minimum safe distance for persons with pacemakers or other electronic implants. This distance will prevent interference of the pacemaker operation by the magnetic field. The 0.1 mT distance is the nominal distance for other equipment that uses video monitors, which minimizes distortion of the image on the monitor by the magnetic field. The manufacturer should be contacted regarding individual situations as the actual distances required are installation and equipment specific.

3.2. GRADIENT SYSTEM

In addition to the shim coil, variations in the main magnetic field are produced through the use of gradients. Gradients are small magnetic fields superimposed on the static field $\mathbf{B_0}$ that vary linearly with distance and are used for spatial localization and first-order field homogeneity correction. Three gradient fields are generated, one each in the x, y, and z directions, using a three-component gradient coil and individual amplifiers or power supplies. Gradient fields are produced by the flow of electrical current through the gradient coil. These fields are applied in two fashions. The primary manner is in short bursts or pulses. The number, duration, and amplitude of the gradient pulses is determined by the particular pulse sequence and measurement parameters. Continuous linear field homogeneity corrections are made using small currents known as gradient offset currents, which are continuously applied through the coil throughout the examination.

The quality of spatial localization that a particular MR system is capable of producing depends primarily on the capabilities of its gradient system. There are four characteristics that are used to assess gradient system performance: maximum gradient strength, duty cycle, rise time or slew rate, and techniques for eddy current compensation. While all four are important for MR imaging, eddy current compensation is the most critical one for MR spectroscopy. Gradient strength is measured in millitesla per meter (mT m^{-1}) or gauss per centimeter (G cm^{-1}) (10 mT m^{-1} = 1 G cm^{-1}), with typical maximum gradient strengths in current high-field MRI systems being 10–25 mT m^{-1}. Larger maximum gradient strengths allow images with greater spatial resolution to be obtained without changing other measurement parameters. The duty cycle of the gradient amplifier is another important measure of gradient performance. The duty cycle determines how fast the amplifier can respond to the demands of a pulse sequence. Duty cycles of 100% at the maximum gradient amplitude are typical for state-of-the-art gradient amplifiers for normal imaging sequences. Large duty cycles allow high amplitude gradient pulses to be used with very short interpulse delays. Gradient rise times are important because the response of a gradient coil to an input current is not instantaneous. The rise time is the time required for the gradient to achieve its final value. The rise time is nominally 0.5–1.0 ms, and can be used to define a rate of gradient change or slew rate. If the desired gradient pulse is 10 mT m^{-1} with a 1-ms rise time, the slew rate is 10 mT m^{-1} ms^{-1}. Shorter rise times and faster slew rates allow shorter gradient pulse ramp durations and interpulse delays within a pulse sequence, enabling shorter echo times. For clinical spectroscopic studies, minimum volume sizes are limited by S/N rather than by gradient limitations. Large tissue volumes ($>10 \times 10 \times 10$ mm³) and low amplitude gradient pulses are typically used, making these performance criteria less important than for imaging studies.

The major complication of gradient pulses for spectroscopic studies is eddy currents. Eddy currents are electric currents produced in response to a changing magnetic field (gradient pulse). They are induced in all metal parts within the MR system, including the body coil and cryoshield, which is the innermost portion of the magnet cryostat. These currents generate magnetic fields that oppose and distort the original gradient pulse. Most eddy currents decay with time constants short compared to the time between the end of the gradient pulse and the beginning of data collection. Those induced in the cryoshield have long time constants and are present during data collection, causing the field homogeneity and the corresponding frequencies to change. Correction of these eddy current induced distortions is known as eddy current compensation. Two approaches are commonly used for compensation. One is to predistort or shape the gradient pulse so that the field variation generated inside the magnet is the desired one. This predistortion may be done via hardware or software. A second approach uses a second coil of opposite polarity that surrounds the main gradient coil, which minimizes the field variation produced outside the main coil, thus minimizing the eddy currents induced in the cryoshield. This approach is called an actively shielded gradient coil. Typical state-of-the-art scanners use both methods of eddy current compensation.

Spectroscopic studies are particularly sensitive to eddy currents. Data acquisition times are relatively long: typically 125–1000 ms. Any field change during this time will distort the spectra. In many instances, additional postacquisition corrections beyond those described above are necessary to obtain well-resolved resonances. The most common approach is to measure the eddy current-induced phase distortion of the localizing gradient pulses to a reference signal. This reference may be a separate acquisition, such as a water-unsupressed spectrum for ^{1}H studies, or a peak within the spectrum, such as residual water. The phase variations of the reference signal are used to remove the gradient-induced distortions from the time-domain signal of the desired volume prior to Fourier transformation.

3.3. RADIOFREQUENCY SYSTEM

The rf system is responsible for generating and broadcasting the rf energy used to excite the spins. It contains four main components: a frequency synthesizer, a digital envelope of rf frequencies, a high power amplifier, and a coil or antenna. Each rf pulse that is broadcast to the patient consists of two parts: a center or carrier frequency and a discrete envelope or function containing a range or bandwidth of frequencies. The frequency synthesizer produces the center or carrier frequency for the pulse. The specific frequency is determined from Eq. 2.5 and is generated as a phase coherent signal by the synthesizer. It is mixed with the rf envelope described below prior to amplification. The frequency synthesizer also controls the relative phase of the transmitter pulse. Most pulse sequences will alternate the phase of the excitation pulse, usually by 180°, to reduce artifacts caused by pulse imperfections. More sophisticated rf systems have synthesizers that allow phase changes of 1–2° increments. This finer control also allows for coherence spoiling through incremental phase change of the transmitter, a process known as rf spoiling.

The rf envelope consists of complex data points, typically 512 in number. These digital points are usually converted to an analog signal prior to mixing with the carrier frequency. Two classes of rf envelopes are used: narrow bandwidth or frequency selective and wide bandwidth or nonselective. Nonselective envelopes, producing rectangular or "hard" pulses, are of short duration and constant amplitude and excite a frequency range of 10–200 kHz with a uniform amplitude at all frequencies. The envelopes are usually used for the determination of the resonant frequency for the spins to be observed from the patient. Nonselective pulses may also be used in a series of pulses applied in a very short time period, known as a composite pulse. The other type of rf pulse is a frequency selective or "soft" pulse. Frequency selective pulses do not have constant amplitude at all frequencies during broadcast. The pulse duration is longer than for a nonselective pulse, allowing for a narrower frequency bandwidth. The frequency bandwidth of the pulse determines the slice thickness and the slice profile. Frequency selective pulses are the most common rf pulses, because they allow the rf excitation to be localized to a specific region of tissue when used with a pulsed field gradient. Chapter 4 describes the types of selective rf pulses used in spectroscopic studies.

The rf power amplifier is responsible for producing sufficient power from the frequency synthesizer signal to excite the spins. The amplifier may be solid state or a tube type. Typical rf amplifiers are rated at 2–10 kW of output power. The actual amount of power required from the amplifier to rotate the spins from equilibrium depends on the specific frequency, coil transmission efficiency, transmitter pulse duration, and desired excitation angle.

The final component of the rf system is the transmitter coil. All MR measurements require a transmitter coil or antenna to broadcast the rf signals. Most MR systems use a saddle design coil to produce uniform rf fields over large volumes (e.g., body or head). This design serves two purposes: to produce uniform rf penetration and to generate an effective B_1 field perpendicular to B_0 even though the coil opening is parallel to B_0. These coils are often adjusted or tuned to the patient to achieve the maximum efficiency in rf transmission. Two types of coil polarity are used, linearly polarized (LP) and circularly polarized (CP) or quadrature. In an LP system, a single coil is present and the rf pulse is broadcast as a plane wave. A plane wave broadcast at a frequency ω_1 has two circularly rotating components, rotating in opposite directions at the same frequency ω_1 (Figure 2.4). For MR, only the component rotating in the same direction at the protons (in-phase) will induce resonance absorption. The other component (out-of-phase) is absorbed by the patient as heat. In a CP transmitter system, two coils are present, one rotated 90° from the other. Equivalent rf pulses are broadcast through each coil. The out-of-phase components cancel each other while the in-phase components add coherently. The patient absorbs only the energy from the in-phase components from each coil so that less rf energy is converted into heat within the patient. A 40% improvement in efficiency from the transmitter system can be achieved for a CP system relative to an equivalent LP system for the same spin rotation (flip angle).

An additional type of transmitter coil is often used for nonproton spectroscopic studies, known as a surface coil. Surface coil transmitters are characterized by a nonuniform B_1 field, and thus a range of flip angles throughout the sensitive volume for the coil. Typically, the flip angle is greatest at locations nearest to the coil surface. However, for planes parallel to the plane of the coil, the B_1 field is relatively homogeneous. This feature has been used, together with the similar homogeneity for surface coil receivers, in the development of localization techniques for MRS (see Appendix A).

While MR is considered a relatively safe imaging technique, the absorbed rf power generates heat inside the patient and the rf amplifiers are capable of generating potentially dangerous levels of rf power. In the United States, manufacturers are required by the Food and Drug Administration (FDA) to monitor the rf power absorbed by the patient so that excessive patient heating does not occur both over the excited tissue volume (localized) and over the entire patient. To accomplish this, the specific absorption rate (SAR) of energy dissipation is monitored. The SAR is measured in watts of energy per kilogram of body weight (W kg^{-1}). The MRI systems are designed to operate at or below the SAR guidelines, which are set to limit the patient heating to approximately 1°C or less. This limit protects tissues with poor blood circulation such as eyes from undue heating that may cause adverse effects. Unlike many imaging studies, most spectroscopic measurements deposit little rf power to the patient and use long TR times, so that SAR limitations are infrequent. One important exception is ^{1}H-decoupled studies, which are particularly rf intensive and may be limited by rf heating concerns.

3.4. DATA ACQUISITION SYSTEM

The data acquisition system is responsible for measuring the signals from the spins and digitizing them for later postprocessing. All MRI systems use receiver coils to detect the induced voltage from the spins following an rf pulse. The exact shape and size of the coil is manufacturer specific, but its effective field, like that of the transmitter coil, must be perpendicular to $\mathbf{B_0}$. The sensitivity of the coil depends on its size and shape. Smaller coils are more sensitive than larger coils, because the patient produces noise that is detected by the coil. For large volume coils such as body coils, the signal is detected from a small region of tissue, but the noise is detected from the entire patient. For small volume coils, such as surface coils, the detected noise is proportionally less. Since the early days of MRI, surface coils have been used as receiver coils because of their high sensitivity to signal originating from a limited region of tissue near the coil and because of their high S/N ratio. The dimension of a coil determines the size of its sensitive volume. Typically, a circular coil with a radius r detects signal strongly from a hemisphere of radius r, with the coil sensitivity greatest for signals produced closest to the coil surface. Newer types of coils known as phased array coils use two or more smaller surface coils to cover a larger area with greater sensitivity than for a single coil of the same size.

The signals produced by the spins are usually nV − μV in amplitude and megahertz (MHz) in frequency. In order to process them, amplification is required, which is usually performed in several stages. The initial amplification is performed using a low-noise, high-gain preamplifier located inside the magnet room or may even be built into the coil itself. The signal is further amplified, demodulated to a kilohertz (kHz) frequency, filtered using a low pass filter, and divided into the real and imaginary parts before being detected by the analog-to-digital converters (ADCs). Each analog signal is digitized by an ADC at a rate determined by the sampling time and number of data points specified by the user. Typical ADCs used on clinical MR systems can digitize a 10-V signal into 16 bits of information at a rate of 5–10 μs per data point. This large dynamic range is of particular importance for ^1H studies where the water concentration is 100,000 times that of the other metabolites. Single-voxel and one-dimensional (1D) spectroscopic measurements do not generate the same amount of data as imaging studies. Raw data set sizes for these studies range from 2–4 Kbytes for a single spectrum to 128 Kbytes for a multivolume data set. In contrast, the raw data for a single image is typically 128 Kbytes. The digitized data will be stored onto a hard disk or onto computer memory for later Fourier transformation or other postprocessing. Phased array coils have a separate preamplifier and pair of ADCs for each coil in the array.

While not formally part of the data acquisition system hardware, an important component of an MRI scanner is rf shielding of the scan room. The weak MR signals must be detected in the presence of background rf signals from local radio and television stations or hospital paging systems. To filter this extraneous noise, MRI scanners are normally enclosed in a copper shield known as a Faraday shield. Maintaining the integrity of this Faraday shield is very important to ensure minimal noise contamination of the final spectra.

3.5. COMPUTER SYSTEM

Every MRI system has a minimum of two computers. The main computer controls the user interface software. This software enables the operator to control all functions of the scanner. Scan parameters may be selected or modified, patient images may be displayed or recorded on film or other media, and postprocessing, such as region-of-interest measurements or magnification, can be performed. Several peripheral devices are attached to the main computer. A hard disk is used to store the patient images immediately following reconstruction. This disk has limited capacity and is used for short-term storage. A device for long-term archival storage, either laser optical disk or magnetic tape, is usually included. A camera will also be attached to the main computer if filming is controlled by the operating software.

One or more consoles will be attached to the main computer. The console is the primary device for operator input. Each console has a keyboard and one or more monitors for displaying images and text information. Many systems use a computer mouse or trackball for more interactive control. Additional consoles may be directly attached to the same main computer or may be separate workstations that access the image data through a network connection.

The second computer that is part of a standard MRI system is an array processor. This processor is a dedicated computer system for performing the multidimensional Fourier transformations to the detected or raw data. The array processor is controlled by the operating software. The raw data is stored from the receiver into memory in the array processor itself or onto a separate hard disk. The array processors currently used with MRI scanners are capable of performing the Fourier transformation for a 256*256 matrix in less than 1 s. Additional array processors may be present which perform computationally intensive postprocessing. For spectroscopic studies, with the exception of three-dimensional (3D) spectroscopic imaging, the limited size of the datasets makes the array processor capabilities less important than for imaging studies.

CHAPTER 4

Techniques for Localized Spectroscopy

4.1. INTRODUCTION

The theoretical development and initial MR spectroscopy studies described in previous chapters were based primarily on pure samples of nonbiological materials. The major clinical applications of MR spectroscopy have focused on the examination of tissues for the purpose of diagnosis of disease or for the monitoring of therapeutic treatments. The in vivo portions of this approach have been facilitated by the use of spatial localization techniques that were originally developed for MR imaging. Spatial localization is the accurate assignment of a detected signal to a specific volume of tissue within an anatomical region of the body. While a multitude of localization schemes have been proposed, the number of techniques that are in clinical use today is limited. A comprehensive review of all localization techniques published to date is beyond the scope of this book. Some of the early approaches to localization are outlined in Appendix A for the reader who is interested in the evolution of these techniques. This chapter describes currently used methods that fulfill the practical restrictions encountered on whole-body scanners in clinical routine practice: relative simplicity, speed, reliability, ease of implementation, and ease of use. Some techniques that are potentially useful for future clinical applications are also reviewed.

Because many of the concepts of spatial localization are used in MR imaging as well as MR spectroscopy, Section 4.2 describes general concepts of localization with magnetic field gradients and with frequency selective pulses. Important issues related to frequency selective rf pulses are outlined; however, no attempt is made at explaining the principles of rf pulse design that require familiarity with complex mathematical and physical theories. Also, a review of basic sequence structure in MR imaging is included to elucidate some of the similarities and differences between MR imaging and spectroscopy.

4.2 BASIC CONCEPTS IN MR LOCALIZATION

Early attempts at localized spectroscopy made use of transmission and reception properties of surface coils. In its simplest form, surface coil localization consists in positioning the region of interest in the sensitive volume of the coil. A nonfrequency selective rf pulse is used to excite all tissues inside this volume. A free induction decay is measured and transformed into a spectrum. Because the effective excitation falls off rapidly outside the sensitive volume, the surface coil itself provides a simple means of localization. However, this concept does not provide very precise localization because the sensitive volume depends on many factors, such as the shape and dimension of the coil and the repetition time of the measurement. Detection of signal from outside the region of interest is difficult to prevent. Furthermore, the coil sensitivity varies significantly across the excited volume, so that signals from some tissue regions are weighted more heavily than others. Nevertheless, surface coils remain a simple and effective tool for localized excitation and detection, and may be combined with other localization methods for better spatial definition and higher S/N.

Current spatial localization procedures confine signals to regions that are considerably smaller than the sensitive volume of a surface coil. These procedures may be used with either volume or surface coils. Localized ^{1}H spectroscopy may be performed with standard coils used for whole-body imaging. Spectroscopy measurements of nuclei other than hydrogen require specially designed transmit/receive coils. Typically, these coils consist of two antennas: one is tuned at the ^{1}H resonant frequency, which is used for imaging, shimming on the water signal, and sometimes for sensitivity enhancing methods such as decoupling and nuclear Overhauser enhancement (NOE)[a]; the other antenna is tuned at the resonant frequency of the nucleus of interest (e.g., ^{31}P or ^{19}F), and is used as both transmitter and receiver for the actual MRS data acquisition.

Localization techniques may be characterized as single volume (or voxel) or multivolume. Single voxel techniques use selective excitation pulses to localize the signal to a relatively small volume of tissue. Using a quadrature head coil, one ^{1}H spectrum can be obtained in about 8 min from a volume of 4 cm^3 or larger. Multivolume techniques can be one-dimensional (1D), two-dimensional (2D), or three-dimensional (3D). The dimensionality refers to the degree of localization, or number of spatial encoding dimensions. In 1D techniques, the MR signal originates from the excited volume, which is a column of tissue divided into slices by gradient or rf encoding. The excited volume can also be the sensitive volume of a surface coil. In 2D techniques,

---

[a]Decoupling and NOE are described in Chapter 2.

the signal comes from a plane that is further sliced into two dimensions by spatially encoding gradient pulses. Individual voxels may or may not contain a mixture of tissues depending on the voxel dimensions relative to those of the region of interest. Some 2D schemes, such as 2D chemical shift imaging, allow a spatial resolution of 1 cm^3 for ^1H spectra, using standard quadrature head coils. Three-dimensional localization excites a large volume, then encodes it in a 3D grid of voxels using gradient or rf pulses. Multiple voxel measurements produce stacked (1D), planar (2D), or 3D arrays of ^1H spectra from adjacent voxels. Because each dimension requires additional phase encoding, longer measurement times are required for multiple voxel techniques, however, the measurement time per voxel is shorter than for a single voxel measurement. Clinical MR spectroscopy employs single voxel spectroscopy (SVS), 2D and 3D chemical shift imaging (CSI), and image selected in vivo spectroscopy (ISIS).

Localization with good spatial resolution requires sophisticated schemes that exploit the physical principles upon which both MR imaging and localized MR spectroscopy are based, namely, the fact that the precession frequency of the magnetization depends on the strength of the external magnetic field. A spatially varying magnetic field, known as a magnetic field gradient, is the key factor in creating the critical correlation between the spatial distribution of spins and the measured signal. Gradients are incorporated into measurement sequences to encode the spatial position of spins by either their precession frequency or their phase. Thus measurement of the frequency and phase distribution over the MR data can be translated into a spatial map of the corresponding spins. In MR imaging, magnetic field gradients are commonly used for encoding three spatial dimensions, for example, as slice selection, phase encoding, and frequency encoding. For reasons explained in Section 4.4, frequency encoding cannot be used in MR spectroscopy.

Figure 4.1 illustrates the superposition of a slice selection gradient, $\mathbf{G_s}$, on a uniform magnetic field $\mathbf{B_0}$. The gradient is applied in the z direction, which causes the spins to experience a magnetic field that depends on their position along the z direction. The field from the slice selection gradient is subtracted from $\mathbf{B_0}$ at negative z locations and added to $\mathbf{B_0}$ at positive z locations. Spins in the negative z region experience a weaker external net field and precess at lower frequencies, while spins in the positive z direction experience a stronger net external field and precess at higher frequencies. In general, the precession frequency of spins at position z is given by a modified Larmor equation:

$$\omega_z = \gamma(B_0 + zG_s) \tag{4.1}$$

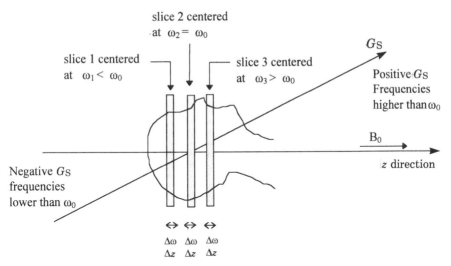

Figure 4.1. A linear magnetic field gradient G_s superimposed over a homogeneous magnetic field B_0 creates a linear frequency variation along the z direction. The center frequency of slice 2, located at the center of the magnet, where $z = 0$ and $G_s = 0$, is not affected by the gradient. Slice 1, in the negative z direction, is centered at a lower frequency ω_1. Slice 3, located in the positive z direction, is centered at a higher frequency ω_3.

The range of frequencies $\Delta\omega$ within a slice having a thickness Δz is

$$\Delta\omega = \gamma \, (G_s) \, (\Delta z) \tag{4.2}$$

where γ is the gyromagnetic ratio and G_s is the amplitude of the magnetic field gradient.

Consider three slices having the same thickness Δz. Slice 2 is in the middle of the magnet where the magnetic field gradient amplitude is zero. It is centered at a resonant frequency ω_2 that would be the same resonant frequency for all spins in a perfectly homogeneous external field B_0 and in the absence of a field gradient ($\omega_2 = \omega_0 = \gamma B_0$). Slice 1 is centered at a lower frequency ω_1 and slice 3 at a higher frequency ω_3. The process of selective excitation makes it possible to measure the MR signal from each slice separately. Selective excitation of a slice is achieved using an rf pulse that has the same center frequency as a given slice (ω_1, ω_2, or ω_3) and that excites the corresponding range of frequencies $\Delta\omega$. The frequency bandwidth of a pulse is inversely proportional to its duration and directly proportional to the slice thickness (Fig. 4.2). The flip angle achieved by a pulse is determined by its duration and its amplitude integral.

Equation 4.2 shows that a given slice thickness can be changed by either adjusting the rf pulse bandwidth $\Delta\omega$ or by changing the gradient amplitude

rf pulse duration   Slice Profile

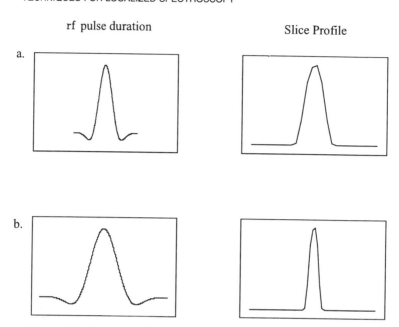

Figure 4.2. The longer rf pulse duration in (b) produces a thinner slice.

G_s. The MR imaging sequences typically use a fixed rf pulse bandwidth and vary the gradient amplitude.[b] Therefore, thinner slices require higher gradient amplitudes. In the early spectroscopy sequences, the rf pulse bandwidth was commonly varied while the gradient amplitude was fixed in order to reduce eddy current distortions to the spectra. Narrower bandwidth pulses are required for thinner slices (Fig. 4.3). However, either method may be used on current state-of-the-art MR scanners.

The excitation profile of an rf pulse determines the accuracy of localization. Ideally, a selective rf pulse should produce uniform excitation of a rectangular slice without exciting spins outside the desired volume. A good slice profile is critical for reducing cross-talk between adjacent slices in imaging and for avoiding spectral contamination with signals originating from outside the voxel of interest in spectroscopy. A rectangular slice profile can only be produced by a selective rf pulse that has a sinc ($=\sin \omega t / \omega t$) envelope and is applied for an infinitely long duration. A sinc function contains a large bandwidth of frequencies. The central lobe of the pulse envelope produces most of the desired flip angle and is determined by low frequencies contained within the pulse. The side lobes, which are defined by high frequencies, determine the square steep sides of the slice profile. For

[b]This allows the TE times and slice profiles to remain constant when the slice thickness is changed.

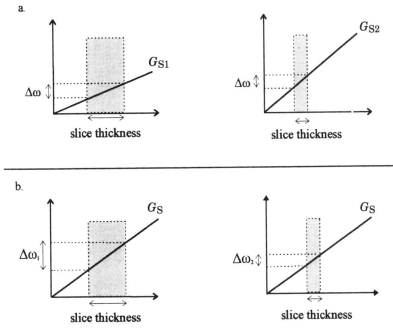

Figure 4.3. (*a*) Fixed rf pulse bandwidth $\Delta\omega$: The steeper gradient G_{s2} defines a thinner slice. (*b*) Fixed Gradient G_s: The narrower rf pulse bandwidth $\Delta\omega_2$ defines a thinner slice.

practical reasons, rf pulse durations are limited to a few milliseconds, typically between 2 and 10 ms. Shortening the duration of the rf pulse is usually accomplished by truncating or eliminating side lobes from the sinc function, which eliminates high frequencies and introduces ripples inside and outside the resulting profile (Fig. 4.4). Multiplying the rf envelope by a filter function reduces the ripples at the expense of wider sloping edges, that is, a broader transition width. For interpretation of MR data, whether images or spectroscopy, it is important to realize that perfectly rectangular slice profiles are technically not feasible. The actual slices or volumes achievable with commonly used rf pulses do not have sharply rectangular edges. The desired width is usually defined as the full width at half maximum height (fwhm) of the slice profile. The implications of slice profile effects on localized spectroscopy are discussed in Section 4.5.2.2.

Filtered sinc pulses, Gaussian, and hyperbolic secant pulses (Fig. 4.5) are available on most clinical imaging systems and are widely used in imaging and spectroscopy applications. Gaussian pulses are characterized by a bell-shaped envelope without side lobes and consequently have a lower rf power deposition than sinc pulses. These pulses produce a better excitation profile for shorter pulse duration; the filtered sinc pulses described above have a

rf pulse duration   Slice Profile

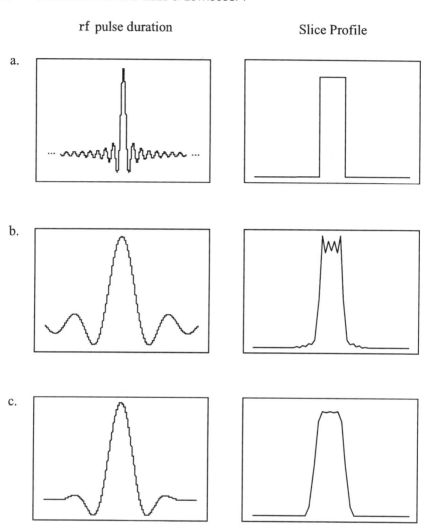

Figure 4.4. Slice profile generated by a sinc pulse. (*a*) An infinite sinc pulse excites a rectangular slice profile. (*b*) A truncated sinc pulse adds ripples to the slice profile. (*c*) A truncated and filtered sinc pulse reduces ripples and widens the transition width or slopes of the slice profile.

more rectangular profile for longer pulse duration. Hyperbolic secant pulses also produce highly selective excitation.[1,2] Above a critical power threshold, a hyperbolic secant pulse creates a sharply defined 180° inversion profile that is independent of the rf power (i.e., **B**$_1$ field). This characteristic makes it highly suitable for frequency selective spin inversion and for surface coil applications. However, the associated rf power deposition is greater than for a sinc inversion pulse. Also, the hyperbolic secant pulse affects only the z

rf pulse duration   Slice Profile

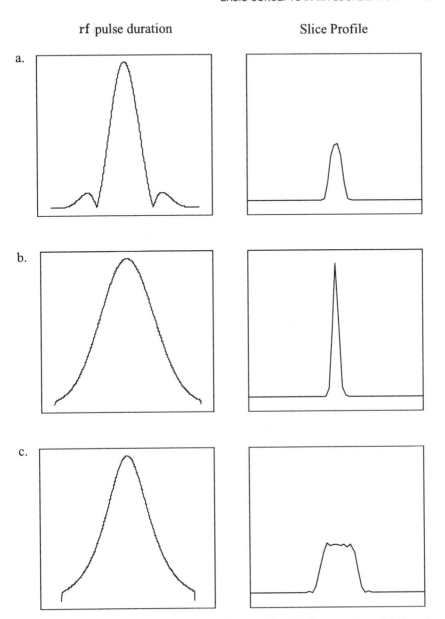

Figure 4.5. Rf pulse envelopes and corresponding profiles. (*a*) Sinc envelope, (*b*) Gaussian envelope, and (*c*) hyperbolic secant envelope.

component of the magnetization and consequently cannot be used as a rephasing pulse, for example, to form spin echoes.

Another class of rf pulses[3,4] have been developed that are frequency selective and $\mathbf{B_1}$ insensitive, that is, they achieve uniform 90° excitation and 180° inversion in the presence of $\mathbf{B_1}$ field inhomogeneities. These pulses are created using frequency modulation that satisfies two constraints. The first one is that decay of the transverse magnetization should be negligible during the pulse. The second is that the rate at which the net magnetization changes orientation must be considerably slower than its rate of precession. For this reason, these pulses are also called adiabatic pulses. The $\mathbf{B_1}$ insensitive, or adiabatic, pulses are especially suited for applications using surface coils as rf transmitters and those requiring uniform solvent suppression or spectral editing.

Most early rf pulse design techniques were based on a relationship between the Fourier transformation of a pulse and its excitation profile. While this approach is strictly valid only for small angles, it works reasonably well for pulses up to 90°. However, it fails for producing good selective 180° refocusing pulses. Current improvements in rf design rely on special computer algorithms for pulse shape analysis and optimization. These algorithms allow for adjustment of a set of parameters that describe the rf pulse envelope until the desired frequency response is generated. Such numerically optimized pulses are the refocusing band-selective, uniform response, pure-phase (reburp) pulses[5] and the Shinnar-Le Roux pulses,[6-11] which have been shown to produce good slice profiles.

4.3. LOCALIZATION IN MR IMAGING

A brief reminder of the structure of an imaging sequence is in order because of similarities in imaging and spectroscopy localization procedures. A detailed explanation of the fundamental principles of MR imaging can be found in Brown et al.[12]

The first step in spatial localization with an imaging sequence consists of exciting a slice of tissue with a frequency selective rf pulse. The pulse is applied simultaneously with a linear field gradient perpendicular to the plane of the slice. Position of the slice is determined by the center frequency of the rf pulse. Its thickness is defined by the gradient amplitude and the frequency bandwidth of the rf pulse. Within the slice, a linear gradient in one direction causes linear variation in precessing frequencies along one dimension. This is called the frequency encoding gradient. Another linear gradient, perpendicular to the frequency encoding and to the slice selection gradients, introduces spatially dependent phase shifts to the precessional motion of the spins. This gradient is referred to as the phase encoding gradient. Within an imaging sequence the MR signal is sampled for multiple values of the phase encoding gradient amplitude, with the same slice selection and frequency encoding gradients. The phase information along with the recorded frequencies are translated into the corresponding spatial distribution in the form of a 2D image (Fig. 4.6).

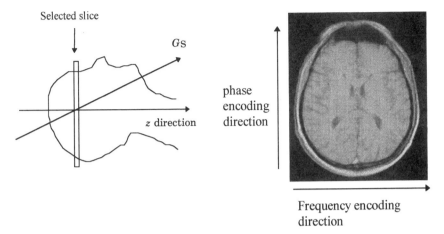

Selected slice

G_S

z direction

phase encoding direction

Frequency encoding direction

Figure 4.6. In MR imaging, one gradient is used for slice selection. The other two are for frequency and phase encoding within the plane of the selected slice. This creates a spatial distribution in the frequency and phase of the MR signal that uniquely identifies each pixel within the slice. The result is a faithful visual reproduction of the various anatomical structures.

A basic spin echo imaging sequence is illustrated in Figure 4.7. The slice selection gradient is on during the frequency-selective rf pulses. The negative lobe refocuses spins that dephased during the long rf pulses. A frequency encoding gradient is applied during sampling of the echo. The additional gradient preceding data sampling refocuses the center of the echo at the center of the ADC window. The gradient following the data sampling period spoils or dephases any remaining transverse magnetization, preventing the formation of unwanted echoes in later sampling periods.

4.4. SPECTRA VS IMAGES

MR images are reconstructed from the entire proton signal, which is dominated by water and fat proton signals. Protons from other metabolites in the body do not contribute to the MR image contrast or S/N because of their negligible concentrations in tissue compared to fat and water protons. In contrast, the goal of MR spectroscopy is to measure those very small metabolite signals, whose resonant frequencies spread over the chemical shift range characteristic of the measured nucleus. For proton spectroscopy, most metabolite signals of interest resonate between the resonant frequencies of water and fat. The water frequency occurs at the high-frequency end (left side) of the range while the fat frequency occurs at the low-frequency end (right side) (Fig. 4.8). In a very homogeneous external field, the resonant

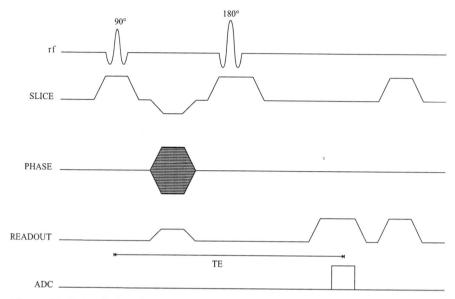

Figure 4.7. Spin echo imaging sequence. Slice selective gradients are applied simultaneously with frequency selective rf pulses. The negative lobe refocuses spins that dephased during the 90° rf pulse. A frequency encoding gradient is applied during sampling of the echo. The additional gradient preceding data sampling refocuses the center of the echo at the center of the ADC window. The gradient following the data sampling period spoils any remaining transverse magnetization, preventing the formation of unwanted echoes in later sampling periods.

frequency of protons in a given metabolite is a measure of their chemical shift, which also uniquely identifies the metabolite peak position. For example, in the long TE normal brain spectrum[c] of Figure 4.8, NAA is identified as the peak positioned at 2.02 ppm. Protons in this NAA peak resonate at a frequency 170 Hz away from the water peak (not shown on this spectrum). The adjoining total creatine (creatine + phosphocreatine or Cr/PCr) and total choline (cho) peaks are separated by 0.2 ppm or 12 Hz at 1.5 T. If the field homogeneity is gradually spoiled, the peaks increase in width until they eventually merge to form one very broad peak. A broad peak would also be produced if a gradient is applied during collection of the MR signal. Therefore in order to preserve optimal homogeneity and chemical shift information, no frequency encoding gradient can be applied during sampling of an MR spectroscopy signal. Thus MRS sequences lack the most rapid form of gradient spatial encoding, namely, the frequency encoding performed by the readout gradient in MR imaging sequences.

[c]The various brain metabolites are described in Chapter 6.

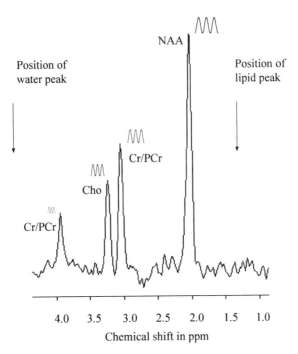

Position of
water peak

Position of
lipid peak

NAA

Cr/PCr

Cho

Cr/PCr

4.0   3.5   3.0   2.5   2.0   1.5   1.0

Chemical shift in ppm

Figure 4.8. Long TE spectrum of normal adult brain at 1.5 T. All the metabolite peaks are within a 3.5 ppm range between the water and lipid peaks, which are not seen here. The arrows indicate positions of the water and lipid peaks at 4.7 and 1.3 ppm, respectively.

As in imaging, localization in spectroscopy is implemented with slice selection and phase encoding gradients. These are applied for a limited time in order to define position and thickness of a slice and to introduce spatial phase encoding to the magnetization. When the gradients are switched off, the spins are back to precessing under the single influence of the homogeneous external magnetic field. Subsequently, the MR spectroscopy signal is sampled without a frequency encoding gradient during data acquisition.

4.5. MRS LOCALIZATION TECHNIQUES

A clinically acceptable MR spectroscopy localization method has to meet several criteria. It must allow definition of small volumes of interest (VOI) with distinct boundaries, using anatomical MR images for reference. It should produce localized spectra from within the VOI with minimal contribution to the signal from outside the defined volume. The spectra must have good S/N with well-resolved, narrow peaks and a relatively clean baseline allowing accurate quantitation. Additionally, the method should be easy and fast to use in routine clinical practice. Localization procedures that meet the criteria for clinical applicability are mostly based on spatial selectivity with B_0 gradients: the same slice selection and phase encoding gradients employed in MR imaging. Spatial localization concepts in current clinical use are developed from

1. Techniques that eliminate signals from outside a volume of interest. These include the use of spatial saturation bands and signal cancellation schemes.
2. Techniques that are based on B_0 gradients like slice selection gradients (DRESS, PRESS, ISIS, STEAM) and phase encoding gradients (CSI) (Table 4.1).

The schemes may be applied either separately or in various combinations, depending on the measurement goals and conditions. These conditions in-

TABLE 4.1. Localization Techniques Used in Clinical Applications

Localization Technique	Localization Scheme	Description: Signal Detection
Surface coil	Coil configuration and positioning with respect to anatomy	Detects signal from sensitive volume of coil
Point resolved spectroscopy (PRESS)	B_o slice selection gradients	Single voxel spectroscopy (SVS) measures signal from one small voxel
Stimulated echo acquisition mode (STEAM)	B_o slice selection gradients	Single voxel spectroscopy (SVS) measures signal from one small voxel
Image selected in vivo spectroscopy (ISIS)	B_o slice selection gradients and inversion rf pulses	• 1D version measures signal from a slice • 2D version measures signal from a small column • 3D version measures signal for a small voxel
Chemical shift imaging (CSI)	B_o phase encoding gradient and B_o slice selection gradient used with optional volume or slice selective excitation	• 1D version measures signal from a slice • 2D version measures signal from a small column (voxel) • 3D version measures signal from a small voxel

clude the anatomical region and the pathology to be examined, coil geometry and performance, metabolites to be observed (concentration, relaxation rates and coupling effects), and the overall duration of a measurement.

4.5.1. Surface Coils

Since the early days of MR imaging, surface coils have been commonly used as receivers because of their high sensitivity. In spectroscopy with nuclei other than hydrogen, surface coils are frequently used as rf transmitters as well as receivers. Their applications evolved from phosphorus spectroscopy studies in small animals[13] to spectroscopy investigations of muscle and superficial organs in humans. Used in a basic pulse and acquire experiment, they suffer from nonuniform rf excitation and inadequate spatial localization. Nonetheless, because of its relative simplicity, good S/N ratio, and short measurement times, localization with small surface coils is still being used in studies of diffuse metabolic effects of superficial homogeneous anatomical structures such as skeletal muscle. Surface coil performance is improved when combined with recent technical developments. These consist of more precise localization techniques, adiabatic pulses (section 4.2) for uniform excitation, and phased array surface coils with a correction algorithm for inherent inhomogeneities in the signal reception profile.[14]

4.5.2. Single Voxel Spectroscopy

Single voxel spectroscopy (SVS) acquires a spectrum from a small volume of tissue defined by the intersection of three orthogonal planes (Fig. 4.9).Two approaches are used for volume definition. The first excites only the volume of interest with frequency selective rf pulses. This approach is used in stimulated echo acquisition mode (STEAM) and point resolved spectroscopy (PRESS) techniques (Fig. 4.10). Both PRESS and STEAM contain a series of three selective rf pulses. However, the flip angles, the sequence timing, and the placement of spoiling gradient pulses are different for the two techniques. The second approach involves excitation and subsequent subtraction of unwanted signals such as in the image selective in vivo spectroscopy (ISIS) technique.

4.5.2.1. Point Resolved Spectroscopy

Point resolved spectroscopy (PRESS)[15,16] is an adaptation of the 1D DRESS[d] technique to localization in three dimensions. The timing diagram of a PRESS sequence is illustrated in Fig. 4.11. With proton spectroscopy,

[d]The DRESS technique is described in Appendix A.

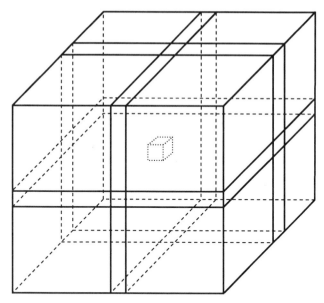

Figure 4.9. A single voxel (interior box) is defined by the intersection of three orthogonal plane.

initial chemical shift selective saturation (CHESS) (Section 4.6) pulses are needed to suppress the water signal. Then, slice-selective rf pulses excite three intersecting orthogonal planes. Slice A (Fig. 4.10) is selected with a gradient G_x and excited by a frequency selective 90° rf pulse. Spins in slice A produce an FID signal, which is ignored. Slice B is defined by gradient G_z and a selective 180° pulse at a time $TE_1/2$ after the 90° pulse. This 180° pulse rephases spins in the intersecting slab, S_{AB} of slice A and slice B. Echo 1 is formed at a time TE_1 after the 90° pulse and $TE_1/2$ after the first 180° pulse. This echo is not sampled. Slice C is selected with gradient G_y and a 180° pulse. Only spins that belong to the intersection of slice C and slab S_{AB} are excited by all three rf pulses. It is their net magnetization that produces the final echo, echo 2, which is sampled and processed. Echo 2 is formed at time $TE_2/2$ from the second 180° pulse. The second G_x gradient is for reversing the dephasing that occurs during excitation. Gradients G_z and G_y are symmetrical about the 180° pulses for complete refocusing of the measured echo.

4.5.2.2. Stimulated Echo Acquisition Mode

Principles of volume selection with stimulated echo acquisition mode (STEAM)[e],[17,18] are the same as with PRESS. Here the three slice selective rf pulses are all 90° pulses (Fig. 4.12). They are applied with slice selection

[e]Section 2.6.1.2 also discusses STEAM.

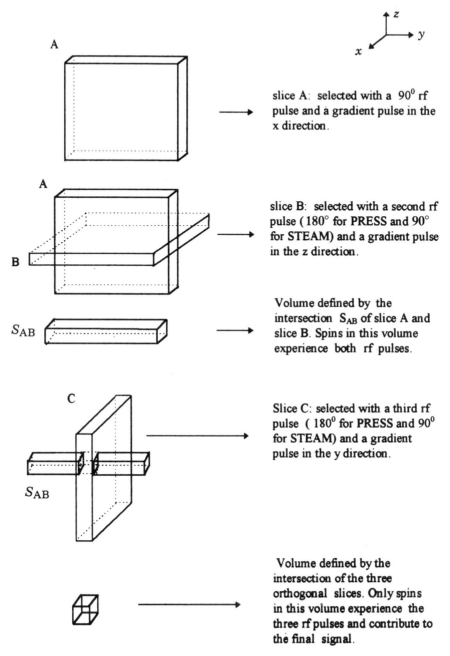

slice A: selected with a 90^0 rf pulse and a gradient pulse in the x direction.

slice B: selected with a second rf pulse (180° for PRESS and 90° for STEAM) and a gradient pulse in the z direction.

Volume defined by the intersection S_{AB} of slice A and slice B. Spins in this volume experience both rf pulses.

Slice C: selected with a third rf pulse (180° for PRESS and 90° for STEAM) and a gradient pulse in the y direction.

Volume defined by the intersection of the three orthogonal slices. Only spins in this volume experience the three rf pulses and contribute to the final signal.

Figure 4.10. Step-by-step excitation process of a single volume with a three pulse PRESS (point resolved spectroscopy, section 4.5.2.1) or STEAM (stimulated echo acquisition mode, section 4.5.2.2) sequence.

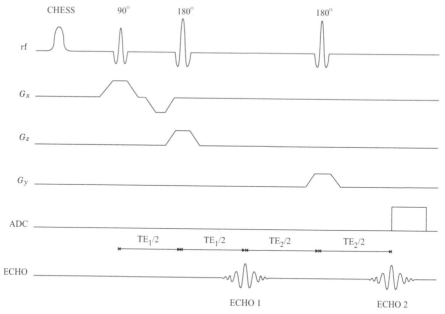

Figure 4.11. Timing diagram of a PRESS sequence: Initial chemical shift selective saturation (CHESS) pulse(s) suppress the water signal. Slice-selective rf pulses excite three intersecting orthogonal planes. The first slice is selected by gradient G_x and excited by a frequency selective 90° rf pulse. Spins in this slice produce a FID signal. The second slice is defined by gradient G_y and a selective 180° pulse at a time $TE_1/2$ after the 90° pulse. This 180° pulse rephases spins in the intersecting slab of the first and second slice, and forms echo 1 at a time TE_1 after the 90° pulse. This echo is not sampled. The third slice is selected with gradient G_y and a 180° pulse. Only spins that belong to the intersection of the three slices experience all three rf pulses and form echo 2 at time $TE_2/2$ from the second 180° pulse. The second G_z gradient is for cancelling the dephasing that occurs during excitation. Gradients G_x and G_y are symmetrical about the 180° pulses for complete refocusing of the measured echo.

gradients each along one of the orthogonal directions x, y or z. The second 90° pulse returns the magnetization to the longitudinal direction where it remains for a time delay TM. The echo time TE is independent of TM. STEAM is derived from an analysis[19] of a three 90° pulse sequence that produces four echoes following the last rf pulse (Fig. 4.13): One echo at TM results from the second and third rf pulses, one echo at TE/2 − TM is formed by the first and second pulses, one echo at TE/2 + TM comes from the first and third pulses and the stimulated echo at TE/2. Gradients in the sequence are carefully set to optimize the stimulated echo and suppress all other echoes and the FID from the last rf pulse, so that only the stimulated echo is sampled during acquisition time. A detailed explanation of gradient optimization with STEAM sequences is found in the literature.[18,20,21] In

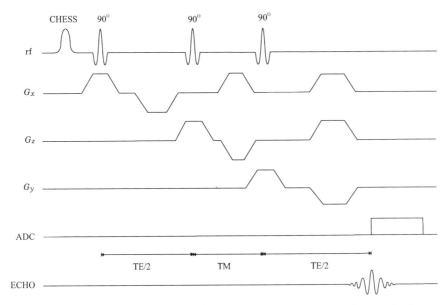

Figure 4.12. Timing diagram of a STEAM sequence: The three slice selective rf pulses are all 90° pulses. They are applied with slice selection gradients each along one of the orthogonal directions x, y or z. The second 90° pulse returns the magnetization to the longitudinal direction where it remains for a time delay TM. The sequence echo time is independent of TM. Gradients in the sequence optimize the stimulated echo and suppress all other signals. Only the stimulated echo is sampled during acquisition time.

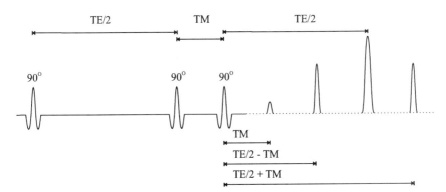

Figure 4.13. Echo formation with STEAM. Three 90° pulses produce four echoes after the last pulse: One echo at TM from the second and third rf pulses, one echo at TE/2 − TM from the first and second pulses, one echo at TE/2 + TM from the first and third pulses and the stimulated echo at TE/2.

proton MRS using STEAM, the water signal may be suppressed using CHESS pulses at the beginning of the sequence[18] and also additional pulses added during the mixing time TM.[21]

The major difference between PRESS and STEAM is in the nature of the echo signal. In PRESS, the echo is formed by refocusing the complete net magnetization, whereas in STEAM, only part of the available signal produces the stimulated echo. Consequently, PRESS has a S/N higher than STEAM, theoretically by a factor of 2. A number of publications have tested and compared the clinical performances of PRESS and STEAM.[22-26] The actual size of the selected VOI for the two techniques is due to pulse profile effects. PRESS uses 180° pulses while STEAM uses 90° pulses. The VOI selection with STEAM yields a volume larger than the one selected with PRESS under the same experimental conditions if sinc pulses are used.[25,26] Better rf pulse envelopes can improve the accuracy of VOI selection. Sequence design considerations, especially with regard to placement of the spoiler and refocusing gradients, make STEAM more sensitive to diffusion than PRESS, particularly if TM is long. However, STEAM also allows for shorter TE, reducing signal losses from T2 relaxation and allowing observation of short T2 metabolites. Also, modulation from J coupling is reduced with short TE STEAM compared to short TE PRESS.

4.5.2.3. Image Selected In Vivo Spectroscopy

The image selected in vivo spectroscopy (ISIS) technique[27,28] consists of a series of measurements having different preparation schemes for the magnetization. The preparation period includes selective 180° inversion pulses that are applied in different combinations prior to collecting an FID from a nonselective 90° pulse. The signals from all measurements are combined algebraically and cancel outside the ROI. ISIS is easily adapted to one-, two- or three-dimensional spatial selectivity.

Figure 4.14 illustrates a 1D ISIS experiment. Here the goal is to obtain a spectrum from slice A. At the start of the experiment, magnetization vectors S_A from slice A and S_{out} from the outside volume are aligned with the external magnetic field. In the first measurement, a 90° nonselective pulse produces a positive signal ($S_1 = S_{out} + S_A$) from the total sensitive volume of the coil consisting of slice A and the outside volume. In the second measurement, a selective 180° inversion pulse is applied simultaneously with a slice selective gradient, prior to the nonselective 90° pulse, in order to invert spins within slice A. Now the signal following the 90° pulse has a negative component ($-S_A$) from slice A and a positive component ($+S_{out}$) from outside the slice. The total signal from the second measurement is $S_2 = S_{out} - S_A$. Subtraction of S_2 from S_1 cancels signal S_{out} and adds the signals from slice A, that is $S_1 - S_2 = 2 S_A$.

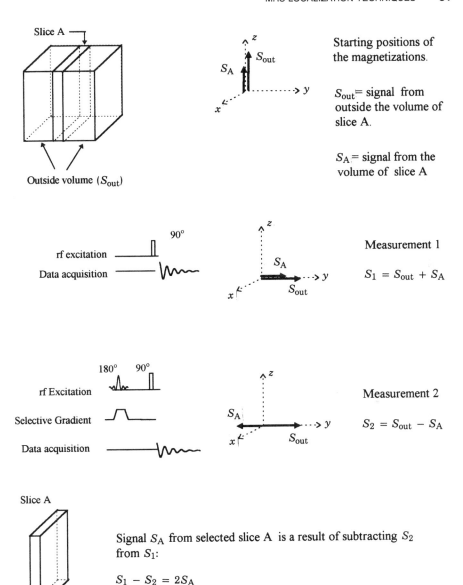

Figure 4.14. One-dimensional ISIS consists of two measurements and localizes the spectrum to a slice.

Two-dimensional ISIS (2D ISIS) (Fig. 4.15) requires four measurements and yields a spectrum originating from the intersecting slab S_{AB} of two orthogonal planes A and B. The signal from the total volume may be thought of as a combination of four terms: the signal of interest S_{AB}; signals S_A and S_B from the remaining volumes of slices A and B, respectively; and signal

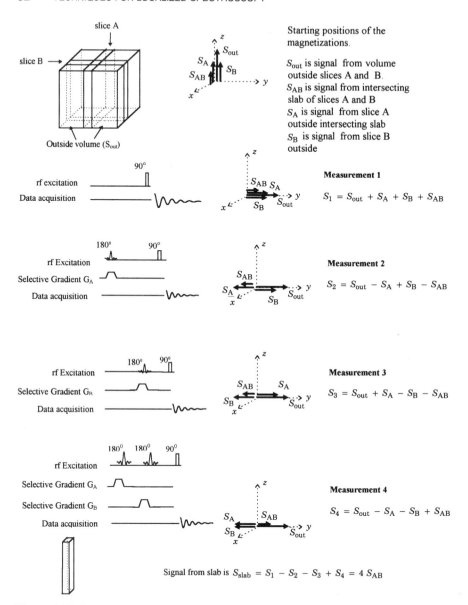

Figure 4.15. Two-dimensional ISIS consists of four measurements and localizes the spectrum to a column.

S_{out} from the rest of the volume. In the first measurement, a 90° nonselective pulse yields an FID from the whole volume. The second and third measurements include one selective 180° pulse and the associated selective gradient perpendicular to either slice A or slice B. The fourth measurement has both inversion pulses and both gradients. Combining the four measurements yields a signal exclusively from the intersection slab of planes A and B.

The 3D version of ISIS (Fig. 4.16) calls for the preparation procedure described above to be repeated in all three directions. A complete 3D acquisition requires eight measurements and yields a spectrum from a small voxel. The various combinations and the resulting signals are listed in Figure 4.16.

The ISIS technique is mostly used for phosphorus spectroscopy, because it can detect fast decaying, that is, short T2*, signals. This is due to the fact that ISIS does not produce an echo but an FID collected right after the rf excitation. The disadvantages of ISIS stem from the fact that it does not measure a signal exclusively from the VOI. Every single measurement includes extraneous signals, which makes it difficult to perform localized shimming on the voxel itself. Signal cancellation from outside the VOI is incomplete if motion occurs in between measurements or if only part of the measurement cycle is acquired. Suggested solutions involve the incorporation of outer volume saturation pulses or alternatively noise pulses that randomize the magnetization everywhere outside the volume of interest. Additional improvement is realized by implementation of 180° hyperbolic secant pulses for optimal inversion of the magnetization. These, along with the noise pulses, are part of a modified version of ISIS called outer volume suppressed image related in vivo spectroscopy (OSIRIS).[29]

4.5.2.4. Advantages and Disadvantages of SVS

As demonstrated in Chapter 6, SVS is currently widely used in a number of clinical applications in the brain. SVS produces a single spectrum from a single localized volume in one measurement sequence. This measurement can take between 2 and 6 min, depending on the volume size, the sequence, and the corresponding parameters. A measurement has to be repeated several times if more than one volume of the anatomy has to be examined. Despite the fact that current technology allows this to be completed in a reasonable time, single voxel techniques are not as efficient as multivoxel techniques that produce spectra from up to 32×32 voxels in one measurement sequence. Nonetheless ^{1}H SVS is gaining popularity, due to its simplicity, ease of implementation, excellent shim readily achievable on small voxels, and immediate access to and interpretation of a single spectrum. Because of its clinical value and clinically acceptable measurement times, some MR centers are making it a routine addition to MR imaging of certain brain pathologies.

4.5.3. Multiple Voxel Techniques

4.5.3.1. Chemical Shift Imaging

Chemical shift imaging (CSI) was conceived in the 1980s as a means for obtaining separate images from either water bound protons or fat bound protons. Currently, CSI is known as a technique that combines features of

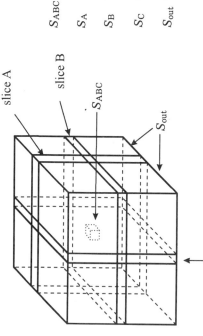

slice A

slice B

S_{ABC}

S_{out}

slice C

S_{ABC} = signal from the voxel at the intersection of slices A, B, and C

S_A = signal from the volume of slice A outside the voxel

S_B = signal from the volume of slice B outside the voxel

S_C = signal from the volume of slice C outside the voxel

S_{out} = signal from outside the volume of slices A, B, and C

Measurement	G_A	G_B	G_C	Signal
1	off	off	off	$S_1 = S_{out} + S_A + S_B + S_C + S_{ABC}$
2	on	off	off	$S_2 = S_{out} - S_A + S_B + S_C - S_{ABC}$
3	off	on	off	$S_3 = S_{out} + S_A - S_B + S_C - S_{ABC}$
4	off	off	on	$S_4 = S_{out} + S_A + S_B - S_C - S_{ABC}$
5	on	on	off	$S_5 = S_{out} - S_A - S_B + S_C + S_{ABC}$
6	on	off	on	$S_6 = S_{out} - S_A + S_B - S_C + S_{ABC}$
7	off	on	on	$S_7 = S_{out} + S_A - S_B - S_C + S_{ABC}$
8	on	on	on	$S_8 = S_{out} - S_A - S_B - S_C - S_{ABC}$

Figure 4.16. Three-dimensional ISIS requires eight measurements and localizes the spectrum to a voxel S_{ABC}. G_A, G_B, and G_C are slice selective gradients perpendicular to slices A, B, and C respectively. They are used simultaneously with 180° inversion pulses. The magnetization in slice A is inverted during measurements 2, 5, 6, and 8. The magnetization in slice B is inverted during measurements 3, 5, 7, and 8. The magnetization in slice C is inverted during measurements 4, 6, 7, and 8. Signal from the voxel is obtained from combining signals from the eight measurements: $S_1 - S_2 - S_3 - S_4 + S_5 + S_6 + S_7 - S_8 = 8 S_{ABC}$.

both imaging and spectroscopy.[30,31] In the literature, it is also referred to as magnetic resonance spectroscopic imaging (MRSI) or simply spectroscopic imaging (SI). Chemical shift imaging is a method for collecting spectroscopic data from multiple adjacent voxels covering a large VOI in a single measurement. Spatial localization is done by phase encoding in one (1D CSI), two (2D CSI), or three dimensions (3D CSI). A CSI sequence is similar to an imaging sequence, but with no readout gradient applied during data collection. Multidimensional Fourier transformation (FT) yields localized spectral data that can be examined in different ways (Fig. 4.17): as single spectra related to individual voxels, as spectral maps, or as metabolite images. Both spectral maps and metabolite images may be overlaid on conventional gray scale MR images.

Processing of 1D CSI data requires two FTs. A spatial FT extracts the spin spatial distribution information and a time FT generates the chemical shift information or spectra. Two conventions are used to describe the dimensionality of CSI techniques (the n in nD CSI). In one convention, n refers to the total number of Fourier transformations that are necessary for data processing. The time domain FT is included as one dimension and 2D CSI, 3D CSI and 4D CSI correspond to spatial localization in one, two and three dimensions, respectively. The other convention uses n to describe the number of spatial dimensions that are phase encoded in the measurement sequence. This convention will be used in subsequent chapters of this book. According to this convention, 1D CSI, 2D CSI, and 3D CSI generate spectra localized to slices, columns, and voxels, respectively (Fig. 4.18). In clinical applications, 1D CSI and 2D CSI are usually combined with other methods, such as slice selective excitation, that localize in the dimensions that are not phase encoded. For example, in 1D CSI, a column of tissue may be defined with slice selective excitation in two dimensions and "sliced" by a phase encoding gradient in the third dimension. If 1D CSI is used with a surface coil, confinement of the volume in the nonencoded dimensions is provided by the sensitive volume of the coil. With 2D CSI, a slice of tissue is defined with a slice selective rf pulse in one dimension and "diced" with phase encoded gradients in the other two dimensions.

Diagrams of nonselective CSI sequences are depicted in Figures 4.19–4.21. A simple 1D CSI sequence (Fig. 4.19), as it is used with surface coils, consists of a nonselective rf pulse followed by a phase encoding gradient. The FID signal is sampled after switching the gradient off and when the spins are back to precessing under the influence of only the external magnetic field. As in an imaging sequence, the experiment is repeated at time intervals TR with incremental values of gradient amplitude. The phase encoding gradient introduces spatially dependent phase shifts. For each TR, there is a different gradient amplitude and a different phase shift distribution

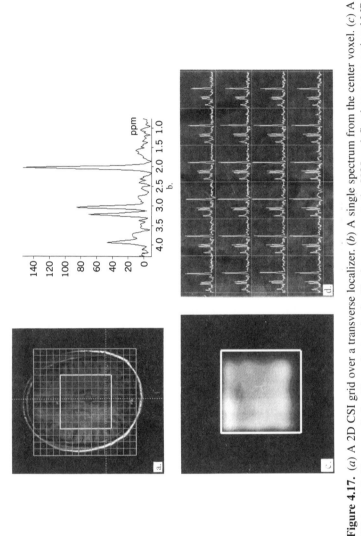

Figure 4.17. (*a*) A 2D CSI grid over a transverse localizer. (*b*) A NAA metabolite map of a healthy volunteer brain. The box represents the VOI as defined on a conventional MR image. The bright area in the box reflects the distribution of NAA within the VOI and illustrates the chemical shift localization offset due to the difference in resonance frequencies between water and NAA. The offset is more pronounced for lower amplitudes of the localization gradients and for a larger chemical shift range. This is analogous to the water-fat chemical shift artifact seen on the conventional MR images. In MRS it is referred to as the chemical shift localization (or offset) artifact. (Chapter 6 contains clinical illustrations of metabolite images). (*d*) Spectral map displaying 24 spectra. (2D CSI data from a brain tumor are shown in figures 6.20, 6.21, and 6.22.

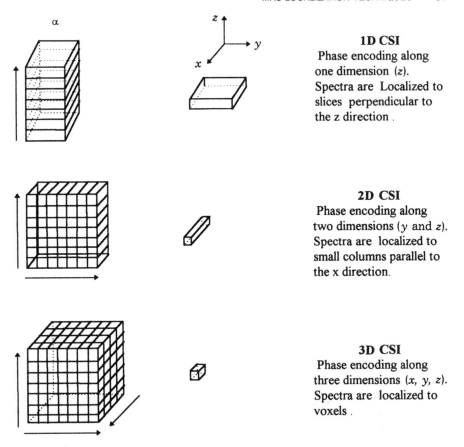

Figure 4.18. Localization to slices, columns, and voxels from 1D, 2D, and 3D CSI.

along the gradient axis. The spatial resolution along this direction is given by the field of view divided by the number of phase encoding steps. A 1D CSI with a 240-mm slab and 16 phase encoding steps produces 16 spectra from 16 parallel and adjacent slices each being 15 mm thick. For a repetition time of 1.5 s and 1 acquisition, the measurement takes 24 s. The total measurement time T_{meas} for 3D CSI is given by

$$T_{\text{meas}} = \text{TR} \cdot N_{\text{acq}} \cdot N_x \cdot N_y \cdot N_z \qquad (4.3)$$

where TR is the repetition time, N_{acq} is the number of acquisitions for each phase encoding step, and N_x, N_y, and N_z are the number of phase encoding steps in the x, y, and z directions. An identical formula with two phase encoding directions applies for 2D CSI. Timing sequences for 2D CSI (Fig. 4.20) and 3D CSI (Fig. 4.21) contain two and three phase encoding tables, respectively, one for each dimension. Phase encoding creates small volumes

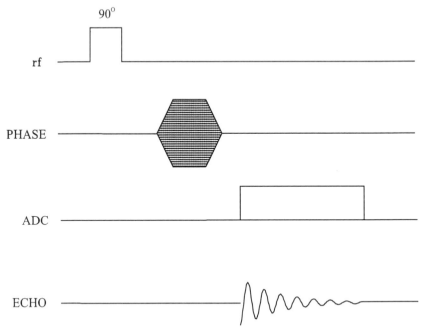

Figure 4.19. Sequence diagram of a nonselective 1D CSI: A nonfrequency selective rf pulse is followed by a single phase encoding table.

arranged in a single layer for 2D CSI or in several adjacent layers for 3D CSI (Fig. 4.18). These arrangements are referred to as 2D and 3D CSI "grids." A 2D CSI measurement having a TR of 1.5 s, 1 acquisition and 16 phase encoding steps in each direction, takes 6 min 24 s. A 3D CSI measurement with comparable parameters takes 102 min. Long scan times and S/N considerations are the reason that 3D CSI is employed with eight phase encoding steps in each direction instead of 16 or 32, thus reducing measurement time while compromising on spatial resolution.

Lower limits on spatial resolution are set by adequate S/N and by a measurement duration that is acceptable for clinical applications. CSI has been used with in vivo ^{1}H MRS in the brain and ^{31}P MRS in brain, muscle, liver, and heart. Proton 2D CSI in the brain is combined with slice selective or volume selective excitation (Section 4.5.3.2) in order to produce voxels as small as 1 cm^3 in about 6–12 min, using a standard imaging CP head coil. Smaller volumes (0.9–0.2 cm^3) have been reported with 3D CSI using a surface coil for high-sensitivity signal reception, and a 19-min measurement time.[32] Because phosphorus has lower sensitivity than protons, ^{31}P CSI requires a typical voxel size of 27 cm^3 for a CP ^{31}P head coil. With phospho-

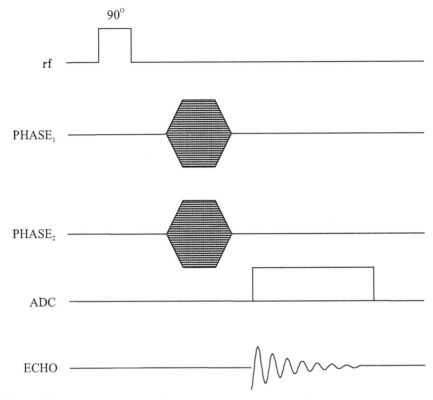

Figure 4.20. Sequence diagram of a nonselective 2D CSI: A nonfrequency selective rf pulse is followed by two phase encoding tables.

rus S/N enhancement techniques such as ^{1}H decoupling[33] and nuclear Overhauser enhancement (NOE),[f] voxel size can be reduced to 10 cm³. [1]H CSI sequences require incorporation of schemes for suppression of water and unwanted lipid signals, which are discussed in Sections 4.6 and 4.7.

The multivoxel capability of CSI is an efficient means for comparing spectra from voxels containing different tissue types. For example, in the case of focal diseases, spectra from a lesion are compared to spectra from normal brain tissue, and heterogeneous metabolic distributions within the lesion may also be investigated. CSI offers a unique advantage in its ability to allow adjustment in voxel positioning after a measurement is finished. This is realized by shifting the CSI grid to the location of interest before calculation of the spatial Fourier transform just as the FOV may be shifted in MRI. This manipulation is called grid shifting or voxel shifting. Another

[f]Section 2.6.1.1 discusses NOE.

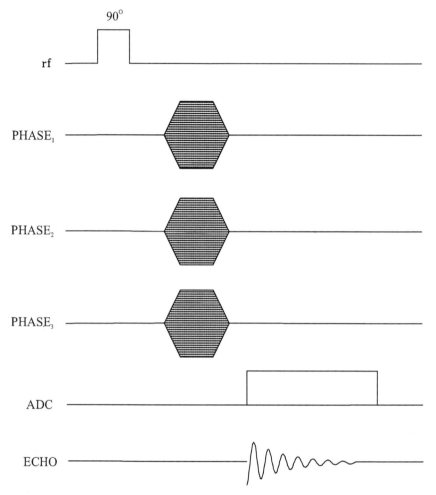

Figure 4.21. Sequence diagram of 3D CSI: A non frequency selective rf pulse is followed by three phase encoding tables.

flexible feature consists of combining adjacent voxels to replicate, as closely as possible, the shape of a lesion and then adding the corresponding spectra.

At present, there are a number of practical problems associated with spectroscopic imaging. One problem is achieving a good shim and uniform water suppression over a large volume that may encompass areas of change in magnetic susceptibility. Another concern is the accuracy of the spatial selectivity that stems from a concept termed *pointspread* function. The pointspread function, which is inherent to all imaging techniques based on Fourier transformation, determines how a signal originating from a point is spread over the reconstructed image. The impact of the pointspread function increases with dimensionality of the spatial Fourier Transform reconstruction. Its implication on spectroscopic imaging is that a spectrum from any given voxel contains contributions from neighboring voxels. The analogous problem in slice-selective techniques such as ISIS or Hadamard[g][34] spatial encoding is the chemical shift artifact. These effects should not be ignored when interpreting spectroscopic imaging data.

4.5.3.2. Volume-Selective 2D CSI

The purpose of volume-selective 2D CSI, or hybrid CSI, is to excite selectively spins within a confined volume that excludes undesired tissues. This is illustrated (Fig. 4.22) by the positioning of a region of interest that is completely enclosed within the skull to avoid exciting the unwanted signals from subcutaneous fat. Volume-selective 2D CSI is commonly used in ^{1}H brain MRS to examine multifocal diseases (Section 6.3).

Principles of volume-selective excitation with CSI are the same as those described for SVS except that the defined volume is normally a large slab. A volume-selective CSI sequence is created from adding to a PRESS or to a STEAM sequence one phase encoding table for 1D CSI, or two phase encoding tables, for 2D CSI. Figure 4.23 illustrates a 2D CSI sequence based on a PRESS sequence. Here phase encoding is done along the x and y directions. All three rf excitation pulses are selective. The z gradient, used for selection of 1.5–2.0 cm thick slices, has the highest amplitude. The x and y gradients define thicker slices (8.0 or 9.0 cm) and have lower amplitudes. Phase encoding covers a field of view that is larger than the excited VOI in order to eliminate possible aliasing of lipid signals. Figure 4.22 is an illustration of the position of an excitation volume (9.0 cm $\times$ 8.0 cm) inside the field of view (16.0 cm $\times$ 16.0 cm) and relative to an axial image of the brain. If the slice is 1.5 cm thick and if 16×16 phase encoding steps are used, the in-plane resolution is 1 cm^2 and the voxel size is 1.5 cm^3 (1 cm $\times$ 1 cm $\times$ 1.5 cm). Signals from this 16×16 matrix are measured simultaneously in one sequence. But only 72 of the 256 spectra are from brain tissue within the excited volume.

Because volume selective proton CSI generates an echo instead of an FID, there is no baseline distortion of the spectra due to the delay between

---

[g]Hadamard spectroscopic imaging (HSI) is a multivoxel technique that uses selective excitation for spatial encoding. As of this writing, HSI has been implemented on clinical MR scanners by several research groups. Because it is not commercially available or commonly used, HSI will not be described in this book.

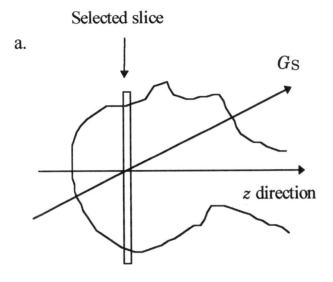

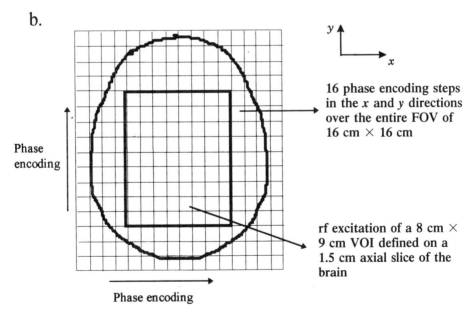

Figure 4.22. Volume selective 2D CSI. (*a*) A slice is selected with the *z* gradient. (*b*) Phase encoding is performed in the *x* and *y* directions over a field of view that is larger than the excited VOI. The VOI is completely inside the brain excluding skull and subcutaneous fat. This set up produces 16 × 16 spectra. Seventy-two spectra are from inside the VOI. Spatial resolution is (1 × 1 × 1.5)cm³.

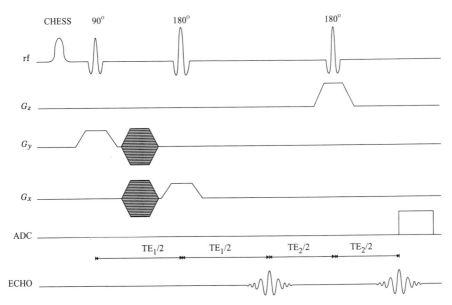

Figure 4.23. Volume selective 2D CSI sequence based on PRESS: All rf pulses are frequency selective in order to excite only a restricted volume. Phase encoding is done along the x and y directions. The z gradient, used for selection of 1.5–2.0 cm thick slices has the highest amplitude. The x and y gradients are for defining thicker slices (8 or 9 cm) and have lower amplitudes.

excitation and signal collection. Typical echo times for 2D CSI are 40–270 ms with PRESS and 20–30 ms with STEAM. Long TE sequences have a flatter baseline because the short T2 lipid signals have decayed at the time when the echo is formed. Short TE sequences allow observations of metabolite peaks not seen with long TE. However, there are multiple poorly resolved resonances from short T2 and J-coupled protons that distort the baseline and complicate peak quantitation. A disadvantage of the single slice volume selective 2D CSI consists of having to repeat a measurement when multiple sections of the brain need to be examined. This prolongs examination time by a factor equal to the total number of measured slices. This problem is solved by the newly developed multiple slice 2D CSI technique that is described in Section 4.5.3.3. Another limitation of volume selective 2D CSI results from enclosing a rectangular box inside the rounded area of the brain. As depicted in Figure 4.22, positioning of the rectangular VOI inevitably keeps out peripheral gray matter thus preventing examination of this area. When this presents a problem for clinical evaluation, it is preferable to use CSI techniques with alternative means for lipid suppression.

4.5.3.3. Multiple Slice CSI

Clinical requirements for multiple slice CSI techniques include thorough examination of pathologies that spread beyond a selected 2D CSI area and adequate brain coverage in the presence of relatively large spatial separation of multiple lesions. In addition, MRS brain screening may be valuable when MR images do not show obvious abnormalities and yet there are clinical reasons for investigating suspected metabolic alterations.

One multisection proton CSI technique[35,36] is based on a standard 90–180° spin echo sequence modified for water suppression with CHESS pulses and for fat suppression with outer volume saturation (Fig. 4.28). Four axial or oblique slices are measured simultaneously by sequential excitation and phase encoding is performed in two dimensions. Measurement time is 27 min for a resolution of 2 cm^3 and an echo time of 270 ms. A variation[37] on this technique uses oblique slices with variable angulation and removes lipid contamination during data processing. Another version of multislice 2D CSI using outer volume saturation employs a gradient echo[38] instead of a spin echo allowing detection of short T2 metabolites. The disadvantage of the outer-volume saturation method is the difficulty of placing fat suppression pulses to cover the subcutaneous fat, since heads deviate considerably from octogonal cylinders.

4.5.3.4. Fast CSI

The most recent innovation in spectroscopic imaging techniques consists in the implementation of fast schemes for diffusion and functional applications in the brain. Functional magnetic resonance imaging (fMRI) and functional magnetic resonance spectroscopic imaging (fMRSI) provide complementary tools for investigation of blood flow and metabolic effects related to brain function. Fast CSI sequences evolved from concepts related to spatial encoding when sampling an MR signal.[39–41] Figure 4.24 is a simplified sketch of an echo planar spectroscopic imaging sequence (EPSI) based on STEAM.[42] The major difference between this sequence and a conventional 2D CSI is that phase encoding is performed only in the *x* direction. The phase encoding gradient table in the *y* direction is replaced with bipolar gradients, switched during data sampling, for spectral-spatial encoding along the *y* direction. Gradient G_z is applied simultaneously with the rf pulses for slice selection. Water is suppressed with CHESS pulses at the beginning of the sequence and in the TM interval. Lipids are suppressed with an outer-volume saturation scheme (Section 4.6). This sequence measures a single 2D CSI slice in the brain with a resolution 0.4–2 cm^3 in 1–16 min. Oscillating gradients and spectral-spatial encoding have been incorporated into 3D spectroscopic imaging sequences.[43,44] A fast 3D CSI measurement takes 34 min for a resolution of less than 1 cm^3. Fast CSI techniques have been demonstrated to work

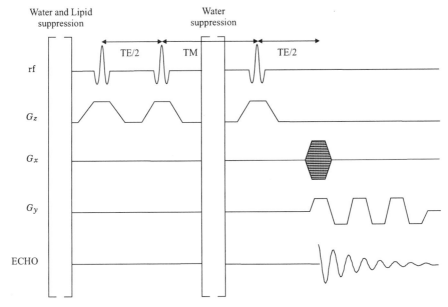

Figure 4.24. A 2D echo planar spectroscopic imaging sequence (EPSI) based on STEAM with schemes for suppressing water and lipid signals. Gradient G_z is applied simultaneously with the rf pulses for slice selection. The major difference between this sequence and a conventional 2D CSI is that phase encoding is performed only in the x direction. The phase encoding gradient table in the y direction is replaced with bipolar gradients, switched during data sampling, for spectral-spatial encoding along the y direction.

on MR clinical scanners. There are problems related to aliasing and to tradeoffs for rapid scan times, such as reduced S/N and more pronounced sensitivity to differences in susceptibility. These are being addressed and the techniques are being tested for their clinical utility. In addition to applications in brain function studies, fast CSI may be a valuable brain MRS screening tool.

4.5.4. Other Approaches to Localization

All techniques discussed so far achieve localization by selecting or phase encoding rectangular volumes that are not representative of anatomical shapes. In addition, the techniques offer limited control over the tissue that may be included in a voxel. This restriction is a serious detriment especially when large voxels are required because of low sensitivity or time considerations. Voxels containing a mixture of tissues add ambiguity to the interpretation of spectral data. To overcome this problem, techniques that allow spectral localization with arbitrary shaped VOIs are being explored.

Spectroscopic imaging with multidimensional pulses for excitation (SIMPLE) (Fig. 4.25) employs multidimensional selective excitation pulses

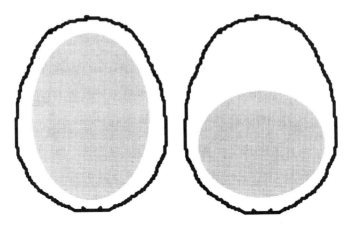

Figure 4.25. Sketch of elliptical VOIs that may be created with the SIMPLE technique.

to define an elliptical volume.[45] This concept allows definition of a VOI that fits inside the skull better than a rectangular volume. The elliptical VOI can easily incorporate peripheral gray matter without interference from undesired subcutaneous fat; two dimensional phase encoding still produces rectangular voxels as in conventional 2D CSI. SIMPLE takes 38 min to collect data from a 1.5 cm thick section using 16 phase encoding steps in each direction and a TR of 1.5 s. The drawback of SIMPLE is that the multidimensional selective pulses have frequency bandwidth limitations that restrict accurate excitation of a wide range of chemical shifts. Therefore, SIMPLE is especially suited for investigations of a single metabolite peak or metabolite peaks with resonant frequencies that are very close on the ppm spectral scale such as creatine and choline. Thorough assessment of all brain metabolite peaks requires multiple measurements and may be a prohibitively long procedure, especially when multiple locations are examined.

Spectral localization by imaging (SLIM)[46] simultaneously measures several spectra from different, arbitrarily shaped, compartments (Fig. 4.26). Prior to the spectroscopy measurement, the compartments are outlined on conventional MR images. The spatial coordinates recorded in this procedure are instrumental to the final processing of the spectral data and to improving the efficiency of the spectroscopy measurement itself. Spectroscopy data are obtained with a modified imaging sequence having a reduced number of phase encoding steps and no frequency encoding gradient during data collection. With SLIM, the minimum number of phase encoding steps needed to spatially encode spectra from N compartments is equal to N. Therefore, measurement of spectra from two different VOIs requires only two phase encoding steps. The reduced number of phase encoding steps translates into a considerable saving of time compared to conventional chemical shift imaging.

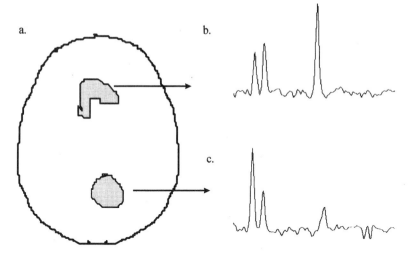

Figure 4.26. A sketch illustrating simultaneous measurement of several spectra from different arbitrarily shaped compartments with the SLIM technique. (*a*) Potential use of SLIM as simulated on a brain lesion. A compartment may be selected within normal brain tissue, another may be within a lesion. (*b*) Spectrum from normal brain tissue. (*c*) Spectrum from lesion.

Advantages to SLIM include short measurement times, no restriction on the shape of a VOI, simultaneous measurement of multiple spectra, and the fact that it can be implemented on standard MR scanners. Because of its speed, it is especially useful for relaxation time measurements where repeated acquisitions are needed to collect data with different values of TE or TR. The major disadvantage of SLIM is that each compartment must be of homogeneous tissue type in order to avoid signal contamination, a condition that is rarely satisfied for clinical MRS.

This problem has been alleviated with a technique called spectral localization with optimal pointspread (SLOOP) function.[47] It was determined that with proper optimization of the phase encoding gradients, spectral contamination from heterogeneous compartments is considerably reduced. The purpose of the optimization is to produce a pointspread[h] function that is constant in phase and intensity within the compartments of interest and that has low intensity outside. This improves S/N and accuracy of the localization. SLOOP has been tested on clinical 1.5 T MR scanners with ^{31}P MRS of human leg[48] and heart.[49] It is showing promise in reducing spectral contamination and measurement time while improving S/N or spatial resolution. The technique requires a special procedure for gradient optimization and a

---

[h]The pointspread function is also discussed in Section 4.5.3.1.

special data processing algorithm that are not currently standard on commercial scanners.

4.6. WATER SUPPRESSION IN PROTON MRS

Clinical MR images are made possible by the strong signal from the inherent high MR sensitivity of protons and by the high water concentration in biological tissue. In MR spectroscopy, sensitivity is critical for detecting signals from small concentrations of metabolites, improving resolution and reducing measurement time. However, the large water signal makes it difficult to observe weak signals from metabolites with concentrations tens of thousands of times smaller than that of water. This is due to a property of the ADC called the dynamic range, which is defined as the ratio of the largest to the smallest signal that can be accurately digitized. It may be that the highest and lowest components in a signal have a ratio that is larger than the dynamic range of the ADC. This results in the smallest signal not being well digitized or even detected. It may also cause "clipping" of the largest signal due to ADC overflow. Clipping the amplitude of a time domain signal introduces artifacts in the form of "fake" peaks in the frequency domain spectrum. Accurate digitization and characterization of weak signals and elimination of "clipping" artifacts are achieved by selectively suppressing the signal with the highest amplitude. This procedure is referred to as solvent suppression, and is commonly used in high-resolution spectroscopy for detection of small signals in the presence of considerably stronger ones.

In clinical ^{1}H MRS, measurement sequences must include ways to suppress the predominant water signal. Suppression does not have to be complete in order to meet the requirement for detection of metabolite signals. On clinical MR scanners, which typically use 16 bit digitizers for a range of 10 V,[i] only a 50–100-fold reduction in the water peak amplitude is needed. The residual water signal may be used during data processing for frequency and phase corrections; it is then subtracted from the MR signal. Elimination of the water signal from the final spectrum by suppression and subtraction, reduces baseline distortions caused by the extended base of the water peak.[j]

Water suppression during measurement is most commonly achieved by spoiling or eliminating the water signal prior to excitation and collection of the metabolite signals. This can be accomplished by water excitation with a

[i]Refer to Chapter 3 for a description of hardware components on a clinical MR scanner.
[j]Refer to Chapter 5 for a discussion of baseline distortions.

frequency selective 90° pulse, known as chemical shift selective saturation (CHESS) pulse, followed by a dephasing gradient pulse.[51] Suppression factors on the order of 100 are achieved by a single CHESS pulse. Multiple repetitions of the selective pulse and dephasing gradient (Fig. 4.27) have been implemented in stimulated echo sequences to produce water suppression factors from 1000[18] to over 10,000.[21] The multiple pulse technique[21] is referred to as DRYSTEAM for drastic reduction of water signal in spectroscopy with the stimulated echo acquisition mode technique. Standard water suppression methods on commercial MR scanners make use of Gaussian narrow bandwidth CHESS pulses. For optimal water suppression, the amplitude of these pulses need to be adjusted prior to a ^{1}H MRS measurement.[k]

The efficacy of water suppression techniques based on CHESS depends on $\mathbf{B_0}$ and $\mathbf{B_1}$ homogeneities, and on the longitudinal relaxation time of water (i.e., T1). Effects of $\mathbf{B_0}$ inhomogeneities may be reduced by optimized shimming. Effects of $\mathbf{B_1}$ inhomogeneities may be eliminated by using adiabatic rf pulses. A recently developed water suppression scheme that is insensitive to water T1 and to $\mathbf{B_1}$ inhomogeneities is described in Ogg et al.[52] The technique is called WET for water suppression enhanced through T1 effects. It uses four frequency selective rf pulses with different flip angles. The advantage of WET is that it produces better water suppression than CHESS without optimization prior to each measurement. System specific optimization of the water suppression pulse amplitudes is usually performed during initial implementation of the technique.

Another approach to water suppression uses inversion recovery of the water signal with a frequency selective 180° pulse. Excitation of the remaining metabolites occurs as the water signal crosses the null point. This technique, called selective water elimination Fourier transform (WEFT) NMR,[53] is not very efficient in cases where values of water T1 vary over a selected VOI. Such variations are common in clinical applications. For example, T1 variations are known to exist between normal tissue and tumors and between gray and white matter. Another disadvantage is waiting for the zero crossing, which prolongs measurement time and TE.

4.7. ELIMINATION OF LIPID SIGNALS IN PROTON MRS

In proton spectra of the brain, peaks are often seen to the right of the NAA signal between 0 and 1 ppm, especially on short TE sequences. These peaks can be either metabolic indicators of lipids or other chemicals from inside

[k]Optimization of water suppression pulses is also discussed in Section 6.3.1.7.

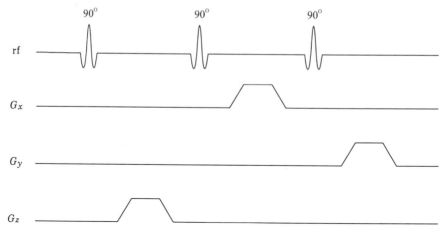

Figure 4.27. Water suppression scheme consisting of a string of three CHESS pulses and spoiling gradients. The amplitude and separation of the pulses are adjusted to produce optimal water suppression.

Figure 4.28. Placement of the outer volume saturation (OVS) slabs for elimination of lipid signals.

the volume of interest, or they may be unwanted signal from lipids that are outside the VOI. The commonly used schemes for eliminating unwanted lipid signals and preserving the metabolic information rely on the selective excitation of a restricted VOI that excludes areas of high lipid concentration or on spatial saturation of external lipids such as those from the scalp and orbital region. Single voxel spectroscopy and volume-selective CSI both employ volume-selective excitation (Sections 4.5.2 and 5.3.2); multislice 2D CSI uses a procedure that surrounds the head with saturation bands carefully placed over the scalp.[36] This technique is referred to as outer volume saturation (OVS) (Fig. 4.28). Alternative means for elimination of lipid signals include data acquisition with long TE sequences and mathematical algorithms that remove lipid signals from the final spectra during data post-processing.[54]

REFERENCES

1. M. S. Silver, R. I. Joseph, and D. I. Hoult: Selective spin inversion in nuclear magnetic resonance and coherent optics through an exact solution of the Bloch-Ricatti equation, *Phys. Rev.* **A31**, 2753–2755, 1985.

2. M. S. Silver, R. I. Joseph, and D. I. Hoult: Highly selective $\pi/2$ and π pulse generation, *J. Magn. Reson.* **59**, 347–351, 1984.

3. K. Ugurbil, M. Garwood, and A. R. Rath: Optimization of modulation functions to improve insensitivity of adiabatic pulses to variations in B_1 Magnitude, *J. Magn. Reson.* **80**, 448–469, 1988.

4. R. S. Staewen, A. J. Johnson, B. D. Ross, T. Parrish, H. Merkle, and M. Garwood: 3-D FLASH imaging using a single surface coil and a new adiabatic pulse, BIR-4, *Invest. Radiol.* **25 (5)**, 559–567, 1990.

5. T. Nagele, U. Klose, and W. Grodd: Numerically optimized rf refocusing pulses in localized MR proton spectroscopy, *Magn. Reson. Imag.* **11(6)**, 785–797, 1993.

6. M. Shinnar, L. Bolinger, and J. S. Leigh: The use of finite impulse response filters in pulse design, *Magn. Reson. Med.* **12**, 75–87, 1989.

7. M. Shinnar, L. Bolinger, and J. S. Leigh: The synthesis of soft pulses with a specified frequency response, *Magn. Reson. Med.* **12**, 88–92, 1989.

8. M. Shinnar and J. S. Leigh: The application of spinors to pulse synthesis and analysis, *Magn. Reson. Med.* **12**, 93–98, 1989.

9. J. Pauly, P. le Roux, D. Nishimura, and A. Macovski: Parameter relations for the Shinnar–Le Roux selective excitation pulse design algorithm, *IEEE Transactions on Medical Imaging,* **10**, 53–65, March 1991.

10. P. Le Roux: Exact Synthesis of radio frequency waveforms, *Proceedings, Seventh Annual Meeting, Society of Magnetic Resonance in Medicine,* San Francisco, 1988, p. 1049.

11. P. LeRoux: Simplified rf Synthesis, *Proceedings,* Eighth Annual Meeting, Society of Magnetic Resonance in Medicine, Amsterdam, 1989, *Works in progress,* p. 1168.

12. M. A. Brown and R. C. Smelka: *MRI Basic Principles and Applications*, Wiley, New York, 1995.

13. J. H. Ackerman, T. H. Grove, G. G. Wong, D. G. Gadian, and G. K. Radda: Mapping of metabolites in whole animals by ^{31}P NMR using surface coils, *Nature (London)* **283**, 167–170, 1980.

14. S. E. Moyher, D. B. Vigneron, and S. J. Nelson: Surface coil MR imaging of the human brain with an analytic reception profile correction, *J. Magn. Reson. Imag.* **5**, 139–144, 1995.

15. P. A. Bottomley: Spatial localization in NMR spectroscopy in vivo, *Ann. N.Y. Acad. Sci.* **508**, 333–348, 1987.

16. R. J. Ordidge, M. R. Bendall, R. E. Gordon, and A. Connelly: Volume selection for in vivo biological spectroscopy, in G. Govil, C. L. Khetrapal, and A. S. Tata, Eds., *Magnetic Resonance in Biology and Medicine*, McGraw-Hill, New Delhi, pp. 387–397, 1985.

17. J. Frahm, H. Bruhn, M. L. Gyngell, K. D. Merboldt, W. Hänicke, and R. Sauter: Localized high-resolution proton NMR spectroscopy using stimulated echoes: Initial applications to human brain in vivo, *Magn. Reson. Med.* **9**, 79–93, 1989.

18. J. Frahm, T. Michaelis, K. D. Merboldt, H. Bruhn, M. L. Gyngell, and W. Hänicke: Improvements in localized proton NMR spectroscopy of human brain: Water suppression, short echo times, and 1 ml resolution, *J. Magn. Reson.* **90**, 464–473, 1990.

19. E. L. Hahn: Spin Echoes, *Phys. Rev.* **80**, 580–594, 1950.

20. W. R. Riddle, S. J. Gibbs, and M. R. Wilcott: Dissecting and implementing steam spectroscopy, *Magn. Reson. Med.* **29**, 378–380, 1993.

21. C. T. W. Moonen and P. C. M. van Zijl: Highly effective water suppression for in vivo proton NMR spectroscopy (DRYSTEAM), *J. Magn. Reson.* **88**, 28–41, 1990.

22. R. Longo and R. Vidmari: In vivo localized ^{1}H NMR spectroscopy: An experimental characterization of the PRESS technique, *Phys. Med. Biol.* **39**, 207–215, 1994.

23. T. Ernst and J. Hennig: Coupling effects in volume selective ^{1}H spectroscopy of major brain metabolites, *Mag. Reson. Med.* **21**, 82–96, 1991.

24. N. M. Yongbi, G. S. Payne, D. J. Collins, and M. O. Leach: Quantification of signal selection efficiency, extra volume suppression and contamination for ISIS, STEAM and PRESS localized ^{1}H NMR spectroscopy using an EEC localization test object, *Phys. Med. Bio.* **40**, 1293–1303, 1995.

25. C. T. W. Moonen, M. von Kienlin, P. C. M. van Zjil, J. Cohen, J. Gillen, P. Daly, and G. Wolf: Comparison of single-shot localization methods (STEAM and PRESS) for in vivo proton NMR spectroscopy, *NMR Biomed.* **2**, 201–208, 1989.

26. L. Kwock, M. A. Brown, and M. Castillo: Extraneous lipid contamination in single volume proton MR spectroscopy: Phantom and human studies, *AJNR* **18(7)**, 1349–1357, 1997.

27. R. J. Ordidge, A. Connely, and J. A. B. Lohman: Image-selected in vivo spectroscopy (ISIS): A new technique for spatially selective NMR spectroscopy, *J. Magn. Reson.* **66**, 283–294, 1986.

28. S. F. Keevil, D. A. Porter, and M. A. Smith: Experimental characterization of the ISIS technique for volume selected NMR spectroscopy, *NMR Biomed.* **5**, 200–208, 1992.

29. A. Connely, C. Counsell, J. A. B. Lohmann, and R. J. Ordidge: Outer volume suppressed image related in vivo spectroscopy (OSIRIS): A high-sensitivity localization technique, *J. Mag. Res.* **78**, 519–525, 1988.

30. T. R. Brown, B. M. Kincaid, and K. Ugurbil: NMR chemical shift imaging in three dimensions, *Proc. Natl. Acad. Sci. USA* **79**, 3523–3526, 1982.

31. A. A. Maudsley, S. K. Hilal, W. P. Perman, and H. E. Simon: Spatially resolved high-resolution spectroscopy by four dimensional NMR, *J. Magn. Reson.* **51**, 147–152, 1983.

32. S. E. Moyher, S. J. Nelson, L. L. Wald, R. G. Henry, J. Kurhanewicz, and D. B. Vigneron: High spatial resolution MRS and segmented MRI to study NAA in cortical grey matter and white matter of the human brain, *Proceedings of the International Society for Magnetic Resonance in Medicine*, Third Scientific Meeting and Exhibition, Nice, France, August 19–25, 1995, p. 332.

33. J. Murphy-Boesch, R. Stoyanova, R. Srinivasan, T. Willard, D. Vigneron, S. Nelson, J. S. Taylor, and T. R. Brown: Proton decoupled 31P chemical shift imaging of the human brain in normal volunteers, *NMR Biomed.* **6**, 173–180, 1993.

34. L. Bolinger and J. S. Leigh: Hadamard spectroscopic imaging (HIS) for multi-volume localization, *J. Magn. Reson.* **80**, 162–167, 1988.

35. D. M. Spielman, J. M. Pauly, A. Macovski, G. H. Glover, and D. R. Enzmann: Lipid-suppressed single and multisection proton spectroscopic imaging of the human brain, *J. Magn. Reson. Imaging* **2**, 253–262, 1992.

36. J. H. Duyn, J. Gillen, G. Sobering, P. C. M. van Zijl, and C. T. W. Moonen: Multisection proton MR spectroscopic imaging, *Radiology* **188**, 277–282, 1993.

37. N. Schuff and M. W. Weiner: Investigation of metabolite changes in cortex and neocortex of healthy elderly by multislice ^{1}H MR spectroscopic imaging, *Proceedings of the International Society for Magnetic Resonance in Medicine*, Fourth Scientific Meeting and Exhibition, New York, April 27–May 3, 1996, p. 1213.

38. K. Maruyama, K. Wicklow, T. Miyazaki, and H. Kolem: [1]H chemical shift imaging of the human brain using a gradient echo, *Proceedings of the International Society for Magnetic Resonance in Medicine, Third Scientific Meeting and Exhibition*, Nice, France, August 19–25, 1995, 1908.

39. P. Mansfield: Spatial mapping of chemical shift in NMR, *Magn. Reson. Med.* **1**, 370–386, 1984.

40. A. Macovski: Volumetric NMR imaging with time varying gradients, *Magn. Reson. Med.* **2(5)**, 479–489, 1985.

41. D. N. Guilfoyle, A. Blamire, B. Chapman, R. J. Ordidge, and P. Mansfield: PEEP- A rapid chemical shift imaging method, *Magn. Reson. Med.* **10**, 282–287, 1989.

42. S. Posse, G. Tedechi, R. Risinger, R. Ogg, and D. Le Bihan: High speed ^{1}H spectroscopic imaging in human brain by echo planar spatial-spectral encoding, *Magn. Reson. Med.* **33**, 34–40, 1995.

43. S. Posse, C. DeCarli, and D. Le Bihan: Three-dimensional echo-planar MR spectroscopic imaging at short echo times in the human brain, *Radiology* **192**, 733–738, 1994.

44. E. Adalsteinsson, P. Irarrazabal, D. M. Spielman, and A. Macovski: Three-dimensional spectroscopic imaging with time-varying gradients, *Magn. Reson. Med.* **33**, 461–466, 1995.

45. D. Spielman, J. Pauly, A. Macovski, and D. Enzmann: Spectroscopic imaging with multidimensional pulses for excitation: SIMPLE, *Magn. Reson. Med.* **19**, 67–84, 1991.

46. X. Hu, D. N. Levin, P. C. Lauterbur, and T. Spraggins: SLIM: Spectral localization by imaging, *Magn. Reson. Med.* **8**, 314–322, 1988.

47. M. Von Kienlin and R. Mejia: Spectral localization with optimal pointspread function, *J. Magn. Reson.* **94**, 268–287, 1991.

48. M. Von Kienlin, J. Hu, H. Jiang, J. M.-Boesh, R. Stoyanova, and T. Brown: Optimized localized spectroscopy of arbitrary volumes on a clinical MR instrument, *Proceedings of the Society for Magnetic Resonance in Medicine*, Twelfth Annual Scientific Meeting, New York, August 14–20, 1993, p. 963.

49. R. Loeffler, M. Von Kienlin, H. Kolem, K. Wicklow, A. Haase, and R. Sauter: ^{31}P spectra of the human heart in vivo with reduced spectral contamination using prior knowledge from MRI, *Proceedings of the Society of Magnetic Resonance,* Second Meeting, San Francisco, CA, August 6–12, 1994, p. 1171.

50. R. Loeffler, H. Kolem, K. Wicklow, A. Haase, and M. Von Kienlin: Localized MR spectra in the human heart from anatomically matched compartments in three spatial dimensions, *Proceedings of the International Society for Magnetic Resonance in Medicine*, Third Scientific meeting and Exhibition, Nice, France, August 19–25, 1995, p. 334.

51. A. Haase, J. Frahm, W. Hanicke, and D. Matthaei: ^{1}H NMR chemical shift selective imaging, *Phys. Med. Biol.* **30(4)**, 341–344, 1985.

52. R. J. Ogg, P. B. Kingsley, and J. S. Taylor: WET, a T1 and B1 insensitive water suppression method for in vivo localized ^{1}H NMR spectroscopy, *J. Magn. Reson. Ser. B* **103**, 1–10, 1994.

53. S. L. Patt and B. D. Sykes: Water eliminated Fourier transform NMR spectroscopy, *J. Chem. Phys.* **56**, 3182–3184, 1972.

54. C. I. Haupt, N. Schuff, M. W. Weiner, and A. A. Maudsley: Removal of lipid artifacts in ^{1}H spectroscopic imaging by data extrapolation, *Magn. Reson. Med.* **35**, 678–687, 1996.

CHAPTER 5

Spectroscopy Data Processing

The MR signal measured from a tissue provides a wealth of information about the biochemicals contained within it. Analysis of this signal may be performed in several ways, but the goals of any analytic method are the following: presentation of MRS data in an easily interpretable format, assignment of measured signals to specific metabolites, and robust determination of the relative or absolute metabolite concentrations. In theory, such information may be extracted from either the measured time domain MR signal or from the processed frequency domain signal.

In the interest of clarity and simplicity, this chapter focuses only on processing operations that are commercially available on clinical MR systems. It explains various processing steps and their effect on the quality of the spectral data. Even though MRS data processing is currently automated on most commercial scanners, an understanding of the underlying principles is valuable for interpretation of spectral features and for detection of artifacts.

5.1. THE MAGNETIC RESONANCE SPECTROSCOPIC SIGNAL

The free induction decay (FID) (Sections 2.3 and 2.4) is a time domain signal that consists of the superposition of several signals that evolve at different frequencies and decay with different time constants T2*. While analysis of the FID itself can be done, it is more common to process the FID using a Fourier transformation. This transformation converts the time domain signal to a frequency domain signal known as the MR spectrum. A spectrum is a display of absorption peaks occurring at the various frequencies that contribute to the signal.

If the MR signal contains a single frequency ω_0 and if the spin system experiences no energy loss, the motion of the resulting magnetization may be represented by a sinusoidal function of constant amplitude, and the corresponding spectrum consists of a spike or delta function positioned at ω_0 (Fig. 5.1). If the same MR signal decays with a time constant T2* due to energy

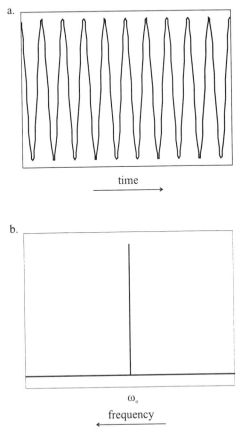

Figure 5.1. (*a*) Sine wave with a frequency ω_0 and a constant amplitude. (*b*) The Fourier transform of the sine wave in (*a*) is a delta function positioned at ω_0.

losses, the corresponding spectrum consists of a Lorentzian peak positioned at ω_0 and having a width that is inversely proportional to T2* (Fig. 5.2).If the MR signal contains several components, each having a different frequency, amplitude, and decay time constant, the FT produces a spectrum with several peaks. The position, amplitude, and width of each peak depend on the frequency, amplitude, and width of the corresponding component (Fig. 5.3).

An example of such a multicomponent FID is the ^{1}H signal from brain tissue. It consists of a large signal from water and several small signals from the various brain metabolites. It has the shape of a large decaying sinusoidal function representing the residual water signal [Fig. 5.4(*a*)] with superposed small "fluctuations" that constitute the sum of small metabolite signals.[a]A

---

[a]For detailed description of ^{1}H brain metabolites, refer to Section 6.3.3.

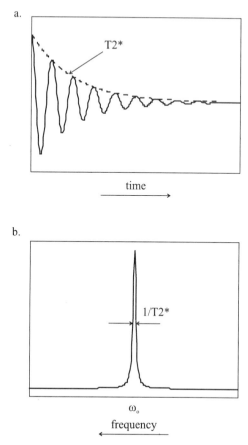

Figure 5.2. (*a*) An exponentially damped sine wave having a frequency ω_0. (*b*) The Fourier transform of the damped sine wave in (*a*) is a peak positioned at ω_0. The width of the peak is determined by the rate of decay T2* of the function in (*a*).

Fourier transformation produces a spectrum that consists of a peak corresponding to the residual water component and very low intensity peaks from all other components [Fig. 5.4(*c*)]. Data processing is necessary in order to remove the residual water signal, to isolate the metabolites signals [Fig. 5.4(*b*)], and to display a useful spectrum [Fig. 5.4(*d*)].

5.2. PROCESSING OF THE MR SIGNAL

While any of the MR signal processing manipulations may be performed prior to or following Fourier transformation, certain operations are computationally easier on one or the other form of the signal. The processing steps that are usually applied to the time domain signal consist of ADC offset

a.

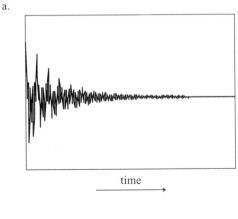

time

b.

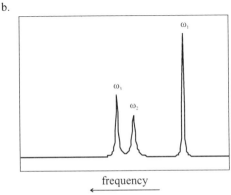

frequency

Figure 5.3. (*a*) An FID made up of three components with different frequencies and amplitudes. (*b*) The corresponding Fourier transform is a spectrum with three peaks having different amplitudes and occurring at the frequencies ω_1, ω_2, ω_3 contained in the FID.

correction, zero filling, and apodization. Other optional processing performed in the time domain are corrections for spectral distortions from residual eddy currents and for removal of the water peak. Following Fourier transformation, phase correction, baseline correction, and curve fitting (including quantification) are performed on the frequency domain spectrum (Fig. 5.5).

During data collection, the analog FID signal is filtered using a low pass filter to eliminate high frequency noise, then goes through an ADC converter (or two filters and ADCs for quadrature detection) where it is sampled[b] with N points for a duration T to produce a digital version of the signal. The time interval t between two sampling points is known as the dwell time and is

[b]In order to avoid complications from high-frequency aliasing, the sampling rate should be at least twice the highest frequency contained in the FID.

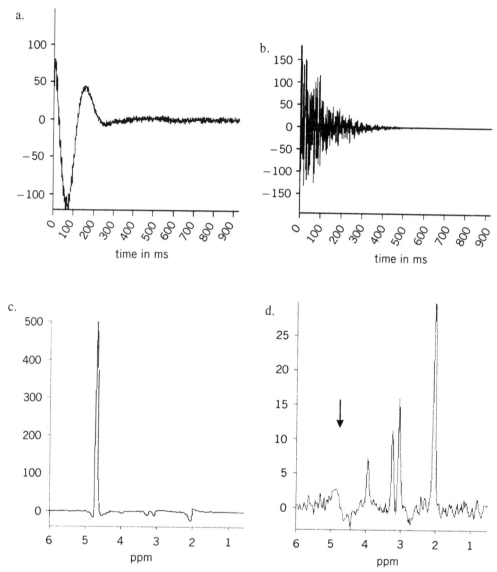

Figure 5.4. (*a*) The signal from a ¹H MRS measurement in the human brain is the sum of a large water signal and small metabolite signals. (*b*) Subtraction of the water signal produces a FID made up of the various metabolite signals. (*c*) The Fourier transform of the FID in (*a*) is a spectrum with a dominant water peak at 4.7 ppm. The metabolite peaks between 1.0 and 4.0 ppm are hardly visible. (*d*) Spectrum obtained from the Fourier transform of the FID in (*b*) and following processing that includes phase correction and baseline correction. This is a typical long TE ¹H MRS spectrum of the normal adult brain. Position of the subtracted water peak at 4.7 ppm is indicated by an arrow.

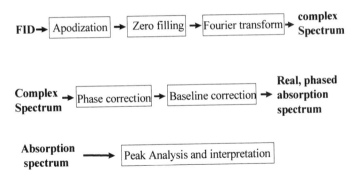

Figure 5.5. Typical sequence of data processing steps for a single spectrum. Following apodization and zero filling, a Fourier transform is applied to the complex time domain data and produces a complex spectrum. The spectrum is corrected for phase and baseline distortions. The real or absorption spectrum is used for peak analysis and interpretation.

given by $t = T/N$ (Fig. 5.6). The Fourier transformation of such a FID is a spectrum having N sampling points at intervals of $1/T$ and a total spectral bandwidth of N/T when using quadrature detection. The digital resolution of the spectrum (in units of Hz per data point) is $1/T$ and determines the frequency resolution within the spectrum. Improvements in spectral resolution may be obtained using longer collection times or measuring more data points. The upper limit for T is usually set empirically and limited by the decay time of the signal or T2*. Extending T significantly beyond the signal decay time results in the collection of background noise, thus degrading the S/N ratio and prolonging measurement time.

The MR signal is usually collected in a quadrature detection mode (Section 2.4) and consists of a complex signal with two components called real and imaginary (Fig. 5.7). Fourier transformation produces a complex spectrum with real and imaginary components that are mixtures of the absorption and dispersion modes (Fig. 5.8) (Section 2.3). The most common spectrum used for analysis is the pure absorption mode, in which the integrated peak area is proportional to the number of spins producing the signal. The magnitude signal, which is a combination of the real and imaginary spectra, is used in the production of MR images. The magnitude signal is phase independent and does not suffer from phase distortions, but it has broader peaks than either the real or the imaginary spectrum. While this does not affect reconstruction and interpretation of conventional MR images, which are derived from large and well-separated water and fat signals, it is not recommended for the analysis and interpretation of spectra that consist of low S/N and sometimes overlapping metabolite peaks.

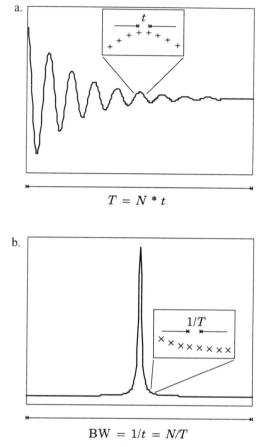

$$T = N * t$$

$$BW = 1/t = N/T$$

Figure 5.6. (*a*) The FID with a duration of $T = Nt$, where N is the number of sampling points and t (shown in inset) is the time interval between two sampling points. (*b*) The frequency bandwidth of the spectrum obtained from the FID in (*a*) is $BW = 1/t = N/T$. The parameter $1/T$ (shown in inset) indicates the spectral resolution, that is, the frequency interval between two sampling points.

5.2.1. Processing of the Time Domain Signal

5.2.1.1. Offset Correction

Based on the theoretical development of MR, the FID from a collection of spins should decay to a mean value of 0 V. If the mean value toward which it decays is nonzero, it is referred to as a DC (for direct current) voltage offset. Any DC voltage produces a spike at zero frequency (transmitter frequency) in the resulting spectrum. The spike may be eliminated during data processing by subtracting the DC offset from the FID prior to Fourier transformation.

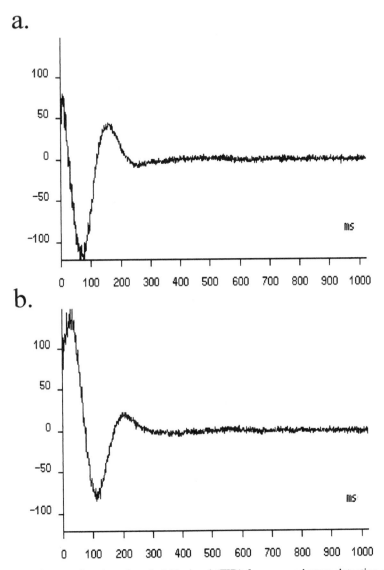

Figure 5.7. The complex time domain MR signal (FID) from a quadrature detection mode consists of (*a*) real part and (*b*) imaginary part.

The most common source of the DC offset voltage can be traced back to the signal detection hardware, namely, "leakage" of the transmitter reference frequency into the receiver electronics. Following signal detection, this yields a background or DC signal that is added to the FID causing it to shift in amplitude. This effect is usually compensated for by adjusting the data collection hardware to produce zero voltage in the absence of an MR signal.

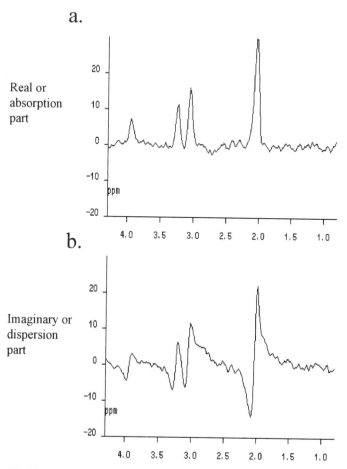

Figure 5.8. The Fourier transform of the complex FID in Figure 5.7 is a complex frequency domain signal or spectrum, which following phase correction is separated into (*a*) a real or absorption part and (*b*) an imaginary or dispersion part.

However, this may not completely eliminate the offset if there is an imbalance in the two quadrature receiver channels. One approach for removing any residual offset, which is often used for measurements using multiple acquisitions, is to alternate the phase of the rf pulse, and thus the phase of the received signal, along with the sign of the ADC during signal averaging. This alternation cancels the constant DC component while adding the signal. If this procedure is not followed during data collection, then a correction will be needed during data processing.

5.2.1.2. Zero Filling
In Section 5.2, it was shown that the spectral bandwidth (BW) is equal to the number of digitization points, N, divided by the data collection time T (i.e.,

BW = N/T). For most MR measurements, the spectral BW is specific to each nucleus and is determined by the range of frequencies contained in the signal. The receiver BW is kept wide enough to detect all spectral frequencies, but narrow enough to minimize the noise contribution to the signal. Period T should be long enough to collect signals from the beginning of the FID but short enough to avoid collecting background noise from the end of the FID. Therefore the number of digitization points of the time domain signal, which is determined by BW and T, cannot be increased arbitrarily in order to increase the spectral resolution. An alternative approach to increasing N in the data collection period is to append additional data points of zero amplitude at the end of the free induction decay following data acquisition. This is called zero filling of the time domain signal. A spectrum acquired with 2048 data points has equivalent frequency resolution to the one acquired with 1024 data points and zero filled to 2048 points (Fig. 5.9). Zero filling the time domain signal achieves the same effect as interpolating the frequency domain signal. The spectral resolution is improved, thereby producing a better representation of fine details and allowing a more accurate definition of the position and height of peaks (Fig. 5.10). Care should be exercised when zero filling an FID that has not completely decayed to noise level. This results in an abrupt change or a step in signal amplitude. Fourier transformation of such a FID produces "wiggles" in the baseline of the resulting spectrum, known as a truncation artifact. In practice, an apodization filter is almost always applied to the FID prior to Fourier transformation; this filter serves both to improve S/N and eliminate truncation artifacts. Apodization causes the FID to decay to zero. Therefore, zero filling an apodized FID does not result in a step-like discontinuity.

5.2.1.3. Apodization

Apodization, also called windowing, is the multiplication of the free induction decay by a filter function to improve either S/N or resolution and/or reduce truncation artifacts. Apodization filtering is different from the low-pass filter mentioned above in that the low-pass filter is a hardware filter applied to the analog signal, while the apodization filter is a numerical multiplication of the digitized time domain signal carried out by software. The most common filter functions decay with time, so that the signal is enhanced at the beginning of the data collection period and is suppressed at the end. This improves S/N at the expense of some broadening of the spectral peaks. A filter function that increases with time enhances the later part of the FID, which improves resolution of the peaks in the spectrum while degrading S/N. Apodization is also used to eliminate truncation artifacts. Multiplication of a truncated[c] FID by a filter function smoothes the

---

[c]A FID is truncated when the data collection period ends before the signal decays to zero.

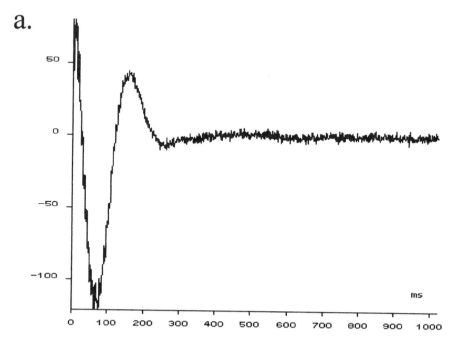

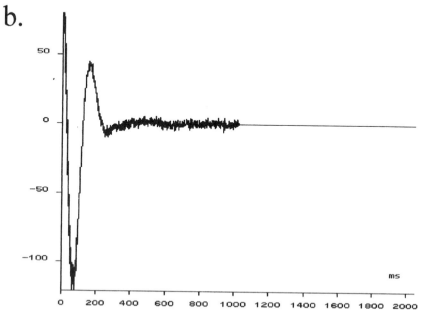

Figure 5.9. (*a*) An FID measured with 1024 sampling points for a duration $T = 1024$ ms. (*b*) Zero filling to 2048 ms consists of adding 1024 sampling points having zero amplitude at the end of the FID.

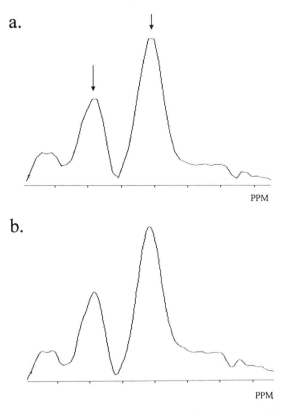

Figure 5.10. Effect of zero filling on the appearance of a spectrum: (*a*) Arrows indicate inadequate sampling of the creatine (3.0 ppm) and choline (3.2 ppm) peaks in a spectrum obtained from an FID that has not been zero filled. (*b*) Zero filling allows better definition of the amplitude and position of peaks.

signal decay to zero, thus eliminating step-like signal discontinuities. This type of discontinuity, if not removed, produces sinc side lobes in the final spectrum. In addition to affecting the spectral line widths, apodization may also alter relative peak intensities, therefore apodization should be applied carefully and specified accurately when reporting results.

Several apodization functions exist, but only a few are employed in clinical MRS and are available on commercial scanners. The most commonly used ones are the exponential (Lorentzian) and the Gaussian functions. An exponential filter is defined by the equation

$$E(t) \sim e^{-t/T_d} \tag{5.1}$$

where t is the time point within the signal and T_d is a parameter that determines the time variation of the filter. The parameter T_d may be positive or negative resulting in a filter that either decays or increases with time. Exponential filters that decrease with time suppress noise from the end of the FID producing an enhancement in S/N while broadening the peaks (Fig. 5.11). Exponential filters that increase with time enhance the latter part of an FID thus prolonging its decay time and, at the same time, raising the noise contribution (Fig. 5.12). The effects on the spectrum are improved resolution with narrower peaks and decreased S/N. The amount of broadening or

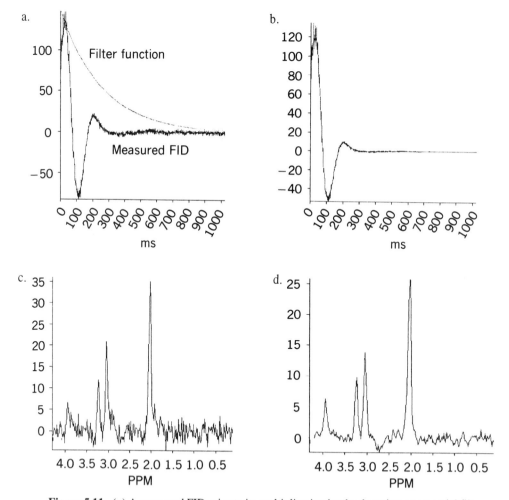

Figure 5.11. (*a*) A measured FID prior to its multiplication by the decaying exponential filter function. (*b*) Result of multiplying the measured FID by the decaying exponential filter function. The filter suppresses noise at the end of the FID. (*c*) Fourier transform of the measured nonfiltered FID with clearly discernible noise around the baseline. (*d*) Fourier transform of the filtered FID. The spectrum has reduced noise and slightly wider peaks.

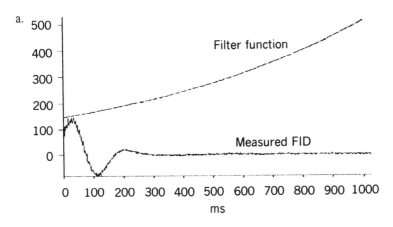

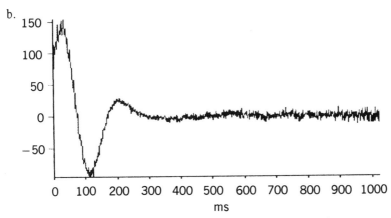

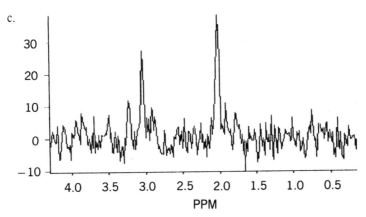

Figure 5.12. (*a*) A measured FID prior to its multiplication by the increasing exponential filter function. (*b*) Result of multiplying the measured FID by the increasing exponential filter function. Noise at the end of the FID is enhanced. (*c*) Fourier transform of the FID in (*b*). The spectrum has increased noise and narrower peaks than the nonfiltered spectrum in Figure 5.11(*c*).

narrowing of the peaks is inversely proportional to T_d. This filter is not commonly used on clinical spectra.

A Gaussian filter is defined by an exponential function decaying as time squared.

$$G(t) \sim \exp[-(t - t_c)^2/T_d] \tag{5.2}$$

Its bell-shaped envelope is controlled by two parameters: the decay time T_d and the center t_c. The decay may be varied to control the amount of spectral line broadening. The center may be shifted such that it coincides with the part of the FID that needs the most enhancement. A Gaussian filter can be used for S/N enhancement or for resolution enhancement with minimal loss in signal amplitude. The Gaussian filter is often described in the literature by a "line-broadening constant" such as "0.2 Hz line broadening." It has been noted[1] that this is an ambiguous description because line broadening caused by a Gaussian filter depends on the original spectral line shape, the original line width, and the filter width (Fig. 5.13).

For optimal signal enhancement, the decay time of a filter should be

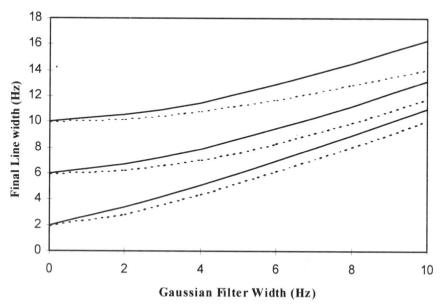

Figure 5.13. Effect of a Gaussian filter on a Lorentzian line shape (solid line) and on a Gaussian line shape (dashed line). Intersections of the curves with the vertical axis correspond to the original line widths of 2, 6, and 10 Hz. The horizontal axis represents the width of the Gaussian filter function, the vertical axis represents the line width after multiplication by a Gaussian filter. [Adapted from Ogg et al. (1995)][1]

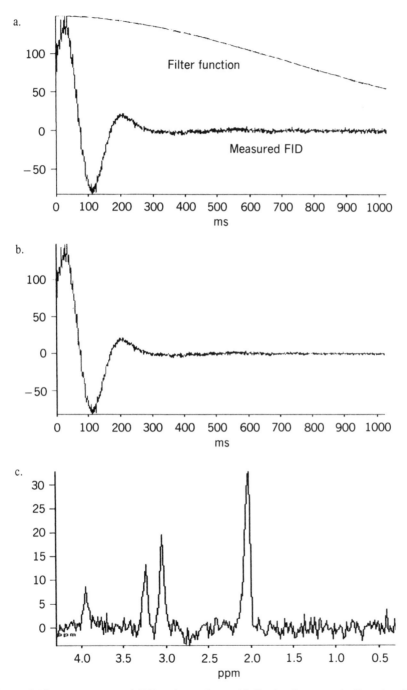

Figure 5.14. (*a*) A measured FID prior to its multiplication by a weak Gaussian filter function. (*b*) Result of multiplying the measured FID by a weak Gaussian filter function. The effect of the filter is hardly noticeable. (*c*) Fourier transform of FID in (*b*).

matched to the decay time of the FID signal (Figs. 5.14–5.16). A filter that decays much slower than the signal is referred to as a weak filter and does not have a noticeable effect on the spectrum (Fig. 5.14). A filter that decays considerably faster than the signal is a strong filter (Fig. 5.16). It causes extreme broadening of the spectral peaks with loss in resolution and in signal intensities. The spectrum in Figure 5.16(c) has a NAA peak broadening of 7 Hz compared to the nonfiltered spectrum. The spectrum in Figure 5.15(c) has a NAA peak broadening of 1 Hz. A broadening of 1 or 2 Hz is usually an acceptable tradeoff for improved S/N and is commonly applied to ^{1}H MRS spectra of the brain (Fig. 5.15). In the case of a multicomponent FID, a filter that matches one of the components may not be optimum for other components. Usually, a compromise is made to improve the overall appearance of a spectrum without significantly distorting any of the peaks.

Another commonly used filter with in vivo MR data is the Lorentz to Gauss filter that converts a Lorentzian line shape to a Gaussian line shape having a narrower peak base. This conversion alleviates the overlapping of peaks and is especially useful for short TE spectra. However, this is appropriate only when the measured line shapes are Lorentzian.

5.2.2. Processing of the Frequency Domain Signal

5.2.2.1. Phase Correction

During a spectroscopy measurement, phase shifts may be introduced as a result of hardware settings or sequence timing. Following Fourier transformation, these shifts introduce a mixture of absorption and dispersion signals into the frequency spectrum. Because quantitative analysis is performed on a pure absorption spectrum (Fig. 5.17), a phase correction procedure is required in order to separate pure absorption and pure dispersion modes into the real and imaginary parts of the complex spectrum, respectively. Most phase shifts encountered in MRS may be corrected with a constant or zero-order phase factor, φ_0, and with a factor that is linearly dependent on the frequency, that is, first-order term, φ_1. For example, a delay between the end of the rf excitation and the beginning of data collection produces frequency-dependent phase shifts. If the delay is t_d, a component of the signal that has a frequency ω experiences a phase shift of ωt_d. Because the various signal

$\longrightarrow$

Figure 5.15. (*a*) A measured FID prior to its multiplication by a medium Gaussian filter function. (*b*) Result of multiplying the measured FID by a medium Gaussian filter function. The filter has no effect on the first part of the FID but suppresses noise from its end. (*c*) Fourier transform of FID in (*b*). Compared to the nonfiltered spectrum in Figure 5.11(*c*), the amount of noise is reduced and the peaks are slightly wider. The NAA peak (at 2.0 ppm) has a broadening of 1 Hz compared to the NAA peak in the nonfiltered spectrum of Figure 5.11(*c*).

a.
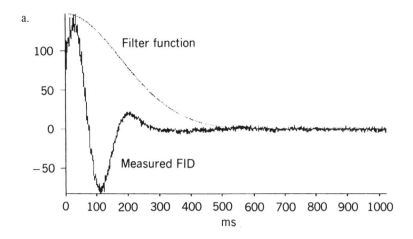

b.
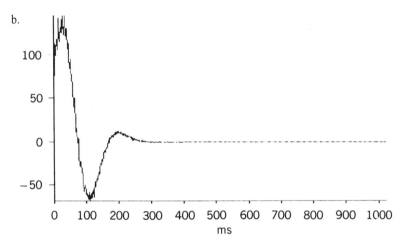

c.
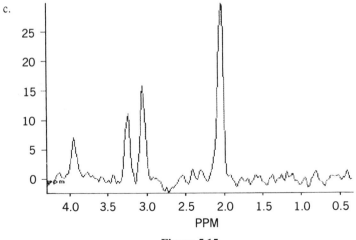

Figure 5.15.

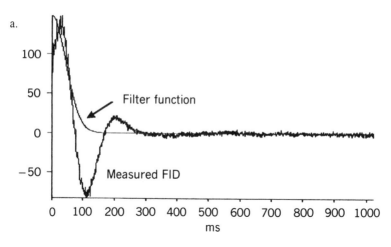

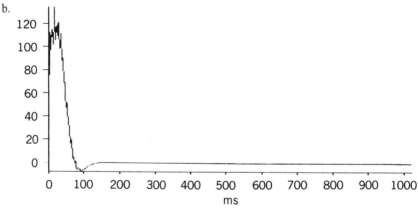

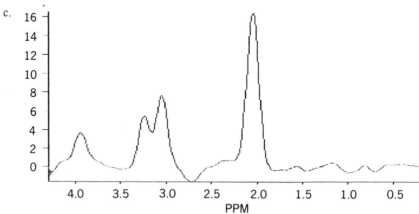

Figure 5.16. (*a*) A measured FID prior to its multiplication by a strong Gaussian filter function. (*b*) Result of multiplying the measured FID by a strong Gaussian filter function. The filter forces a rapid decay of the FID. It suppresses noise from the end of the FID and reduces the signal at the beginning of the FID. (*c*) Fourier transform of FID in (*b*). A strong filter causes wider spectral peaks and a loss in resolution as seen on the Cre and Cho peaks (at 3.0 and 3.2 ppm, respectively). The NAA peak (at 2.0 ppm) has a broadening of 7 Hz compared to the NAA peak in the nonfiltered spectrum of Figure 5.11(*c*).

124

components have different frequencies, they undergo different phase shifts. This type of phase shift requires a first-order phase correction, φ_1. On the other hand, if the phase of a quadrature detector is incorrectly adjusted, or if there is a fixed phase difference between the transmitted and received rf signal, all compnents experience the same amount of phase shift. In these cases, a zero-order phase correction φ_0 is sufficient.

Phase correction is performed mathematically by multiplying the signal by a factor of the form $\exp[i(\varphi_0 + \omega\varphi_1)]$, where φ_0 is the constant phase term and φ is the frequency-dependent phase term. The two factors are varied independently, either manually or automatically, until the best separation of the absorption and dispersion modes is achieved.

5.2.2.2. Baseline Correction

Under ideal circumstances, the baseline of an MR spectrum will be flat until the resonance peak is reached. There are several factors that may introduce distortions to the baseline. A delay between the rf excitation and the beginning of the collection period produces a rolling baseline, even after appropriate phase correction for the delay. This distortion may occur with a CSI sequence where time should be allowed for phase encoding before collecting an FID signal. Other baseline distortions are due to broad spectral humps that are signals from immobile nuclei (short T2*) such as ^{31}P phospholipid nuclei in phosphorus MRS or macromolecules in ^1H MRS. With ^1H MRS, if a residual water signal is not subtracted during postprocessing, the edges of the water peak will cause sloping of the baseline, especially on short TE spectra [Fig. 5.18(a)].

For reliable evaluation of peak areas, all distortions must be eliminated and the baseline must be as flat or as well defined as possible. A commonly used procedure is to fit sections of the baseline that are in between peaks to a curve that is subsequently subtracted from the spectrum. Care must be taken with baseline correction due to the possible distortion of signal intensities; Figure 5.18(c) is an example of such distortions. Lorentzian peaks have a large fraction of the total area in the extended "wings" of the peak, and are thus subject to errors in determining the peak area if the baseline is incorrectly defined.

5.3. INTERPRETATION OF SPECTRA AND QUANTITATIVE ANALYSIS

Interpretation of MRS spectra consists of identification of the various spectral peaks, calculation of the relative or absolute metabolite concentrations, and determination of quantities from peak shifts, such as intracellular pH or

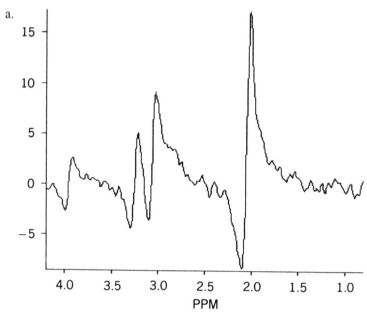

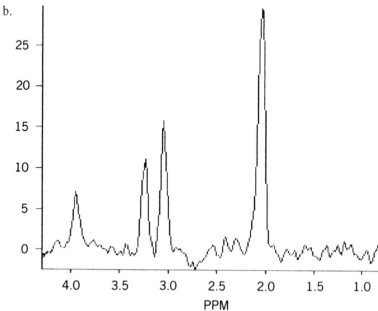

Figure 5.17. (*a*) A long TE ¹H MRS normal adult brain spectrum (TE = 135 ms) with phase distortions. (*b*) Pure absorption spectrum obtained from correcting the phase of spectrum (*a*).

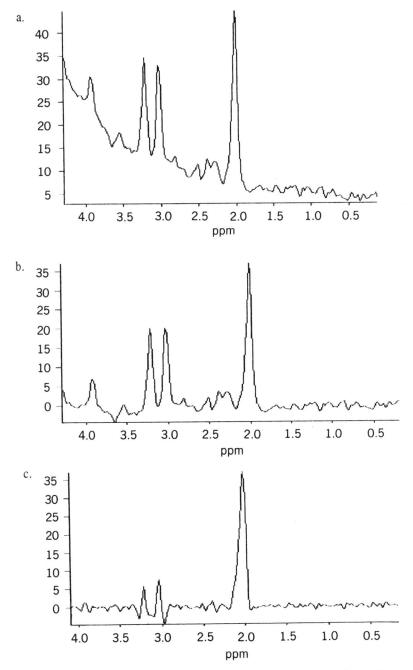

Figure 5.18. (*a*) Baseline distortion of an ¹H MRS brain spectrum due to incomplete water suppression. The metabolite peaks are seen on the decaying "wing" of the water peak. (*b*) Baseline correction with minimal effects on the spectral peak. (*c*) Example of peak distortions introduced by inaccurate baseline correction.

magnesium ion concentration. In many situations, qualitative data interpretation may quickly provide the needed answers. However, it is the ability to assess in vivo metabolic levels that gives MRS a clear advantage over other clinical imaging techniques. Accurate and reliable quantitation, when achieved and routinely applied, will have the biggest impact on the clinical utility of MRS.

5.3.1. Peak Identification

The first step in any MRS data evaluation is identification of the various spectral peaks and their assignment to particular metabolites. The initial and formidable task of peak assignment relies on readily available chemical shift information from high-resolution MR spectroscopy. Signals from most of the major metabolites visible at 1.5 T in human tissues have been identified and more are still being investigated. Verification of a given peak assignment is performed by in vitro measurements on a reference solution containing the metabolite in question under near physiological conditions. Another method for verifying peak assignment is to measure high-resolution spectra from extracts of the tissue of interest.

The collected peak assignment data have been published and are available[2,3] to aid in the interpretation of clinical MRS spectra.[d] Complete peak identification requires determination of peak positions and characteristics. For example, in ¹H MRS of the brain, it is known that lactate and a signal from the methylene protons of mobile lipids both occur at the same position, that is, 1.3 ppm; therefore signal identification needs other discriminating factors, such as peak characteristics. It is known that mobile lipids have two separate peaks, one at 0.9 ppm from the methyl protons, the other at 1.3 ppm from methylene protons. Lactate has a characteristic doublet, a combination of two peaks separated by a *J* coupling of 7 Hz. It is also known that the lactate doublet is inverted at a TE of 135 ms.[e] Matching this type of information to the signal observed at 1.3 ppm helps in its assignment either to lactate or to lipids or a combination of both.

Peak identification alone may be useful in instances where only verification of the presence or absence of a given metabolite is needed, or when a discriminating piece of information may help with controversial MRI results. For example, MRS is routinely applied for verifying the presence or absence of lactate as a result of pathology, relying on the fact that significant

---

[d]Refer to Table 6.3 for a listing of brain metabolites and the corresponding peak positions.
[e]Although TE = 135 ms has been commonly used in the literature, the theoretically correct value for inversion of the lactate doublet is TE = 144 ms. Refer to Section 2.6.1.2 for a discussion of the lactate doublet characteristics.

lactate levels in brain tissue is abnormal. An inconclusive MRI exam of a brain lesion may benefit from an ^{1}H MRS spectrum. An obviously absent NAA peak helps confirm suspicion of damage or loss of neurons or axons. Nonetheless, in most clinical situations, some type of quantification is essential for assessing metabolite levels in patients compared to those in healthy individuals, and for monitoring changes in metabolite levels of patients undergoing treatment. A major factor in the acceptance of in vivo MRS as a useful clinical tool relies on its ability to provide an accurate and reproducible means for determining metabolite levels.

Peak identification and quantitation are affected by the measurement sequence and parameters. The number of observed peaks, their amplitude, intensity, and phase (Section 6.3) depend on TR and TE compared to the relaxation time of the metabolites. Short TE spectra contain more peaks due to contributions from short T2 components. While more information is contained in these spectra, they are more difficult to interpret. Long TE sequences are sometimes used to simplify spectra by reducing the contribution from short T2 signals.

5.3.2. Peak Editing

Peak editing, or spectral editing, is a means of separating overlapping peaks in order to facilitate their identification. It also brings out hidden peaks that are not seen otherwise. Editing techniques exploit differences in properties of the metabolites and their peak characteristics such as relaxation times or spin coupling. Most editing techniques prolong MRS exams because they require additional measurements. Few techniques fulfill the requirements for clinical in vivo MRS: relative simplicity, speed, and ease of implementation on standard MRI scanners. In the example of the lipid and lactate peaks mentioned above, sophisticated editing techniques for lactate detection have been devised in research laboratories. Nonetheless, in clinical interpretation of data, editing relies on basic concepts related to the fact that the lactate signal is a doublet, and that it is inverted at TE = 135 ms and upright at TE = 270 ms. Editing by inversion recovery is a technique based on differences in metabolite T1 relaxation times. It employs a frequency-selective 180° rf pulse to invert only the spin species that need to be suppressed. The signal from other species is collected as the magnetization from the inverted spins crosses zero on its way to recovery. Inversion recovery is commonly employed in MR for suppression of water and fat. Other editing methods encountered in vivo include T2 dephasing with long TE sequences, and selective excitation and spoiling, such as water suppression CHESS pulses in ^{1}H MRS (Section 4.6).

5.3.3. Peak Area

5.3.3.1. Peak Area Calculation

The use of MRS for quantitative analysis depends on the fact that, if certain conditions are met, the area under a peak is directly proportional to the number of spins contributing to the peak:

$$\text{Peak area} = K \text{ (number of spins)} \qquad (5.3)$$

Therefore, under appropriate conditions in MRS evaluation, assessment of metabolite relative or absolute concentrations can be reduced to calculation of peak areas, also referred to as peak integrals.

Several approaches have been used for computing peak areas. The first and simplest way is to measure peak heights. This approach is precise and accurate only if the peak height is directly proportional to the peak area, which is true only in spectra where peaks can be guaranteed to have the same line shape and the same line width in all evaluations. This condition does not hold in clinical MRS. The peak height measurements must be regarded as unreliable indications and used for rough approximations only. Another method called triangulation consists in drawing a triangle that best fits a peak and calculating its area. Triangulation has the obvious drawback of being approximate and operator dependent, and it is not recommended for the low-field broad line spectra of clinical MRS.

The minimally adequate peak quantitation method is based on numerical integration. A common approach uses software programs that allow a peak to be defined by selecting the integration frequency range with the help of a cursor. In addition to being operator dependent, errors are introduced from uncertainties in determining the extent of peak wings and defining the base-line, which is most problematic with Lorentzian line shapes. This problem may be somewhat alleviated by converting the Lorentzian lines to Gaussian lines with a Lorentz-to-Gauss transformation. Another complication occurs with overlapping peaks, which may be difficult to define and evaluate separately and may have to be evaluated as a single peak. Numerical integration is easy to apply on high S/N spectra with well-resolved peaks and a flat baseline; but these are not conditions that apply to clinical MRS. Although still in use today, numerical integration is being replaced by more sophisticated spectral analysis programs.

A more refined approach to peak area calculation requires curve fitting. This technique uses multiple regression analysis to find the best match between a theoretical curve and the measured spectrum. Some prior information is required to make a good initial estimate of the peak parameters, which may be provided by the operator or by the program. For example, this information might include peak line shape functions (Lorentzian, Gaussian, or a mixture of the two[4]), the frequency range to be fitted, and approximate values for the line widths, peak positions, and peak heights. Then an iterative process is performed to obtain the best fit between the theoretical and measured curves. The advantage of this method is that it yields more accurate results for the overall spectral evaluation, especially where there are overlapping peaks. Its drawbacks are that the actual line shapes in the measured spectrum may not be those assumed by the software and therefore the best fitted theoretical curve may not exactly represent the true measured line shape. Spectral curve fitting is widely used and accepted in reporting clinical MRS results. It is a technique that is currently implemented on most clinical MR scanners.

Curve fitting techniques may be applied to the time domain signal instead of the processed spectrum, thus fitting the originally acquired data that have not been subjected to manipulations that affect the width, amplitude and area of peaks. Time domain curve fitting techniques are better suited for analysis of truncated signals. One technique,[5] the least prediction single value decomposition (LPSVD), has proven to be efficient at separating signal components that produce overlapping peaks in the frequency domain signal. However, like the previously mentioned approaches, it assumes line shapes that might not be a good representation of the measured signal.

There are several advances in automatic signal processing techniques[6–9] that minimize preliminary assumptions and reduce operator bias and other factors that may influence the outcome of the analysis. For example, one approach[6,7] models the MRS data as a combination of signal (peaks), a rapidly varying component or noise, and a slowly varying component or baseline. Another approach is based on the time domain signal analysis using the Bayesian statistical technique[8] for reconstruction of the spectrum and for derivation of the various peak parameters with minimal prior assumptions. The Bayesian analysis is presently extremely computer intensive for clinical use. These techniques are being tested at research centers for clinical applicability, and are not currently available on clinical MR units. Regardless of which data processing method is employed, optimizing the quality of the measured raw data is crucial to obtaining reliable and accurate spectral data. Whenever possible, it is important to minimize or completely eliminate sources of artifacts from the data collection procedure (Section 6.3.2).

5.3.3.2. Normalized Peak Area

Peak area calculation by itself is not sufficient for accurate assessment of concentrations because the proportionality constant K in Eq. 5.3 is unknown. Area calculation may be used in three ways: to derive relative deviations from normal standard values; to derive relative concentrations of metabolites from ratios of peak areas; or to determine absolute concentrations using a correction for relaxation (T1 and T2) effects and prior knowledge about "NMR invisibility."[f]

With any quantitation method, factors that affect peak areas must be taken into account. Therefore normalized peak areas are derived from the calculated peak areas and the appropriate correction factors. These factors may be either characteristic of the examined tissue such as relaxation time effects, f(T1, T2), or experimental, f(exp), such as partial volume effects from rather large voxels, differences in coil loading and in voxel size and possible $\mathbf{B_0}$ and $\mathbf{B_1}$ field inhomogeneities.

$$\text{Normalized peak area} = (\text{calculated peak area}) \, (f(T_1, T_2) \, (f(\exp)) \quad (5.4)$$

Normalized peak areas are used in comparing spectra from various patients and normal volunteers and in evaluating changes in spectra from serial examinations of the same patient. Accurate interpretation of such data depends on the understanding of the various correction factors and their effect on peak area calculations.

Relaxation time effects consist of signal loss from T1 saturation due to short repetition times and from T2 dephasing, especially on long TE measurements. Corrections for these effects require knowledge of the metabolite relaxation times, T1 and T2, in the tissue of interest. Measuring T1 and T2 on patients is not feasible and the use of relaxation time measurements from normal volunteers (Tables 5.1 and 5.2) do not apply to patients because tissue relaxation times are different in pathological tissue. Alternatively, signal loss from relaxation time effects may be minimized by careful selection of the measurement parameters. T2 dephasing is reduced with the use of short TE sequences and signal saturation is avoided with the use of a TR that is long enough to allow almost full relaxation of the metabolite with the longest T1. Typical TE values are 20–30 ms[g] for quantitation of ^{1}H brain spectra, and less than 10 ms for ^{31}P spectra. Typical TR values are about 8 s for protons and 10 s or more for phosphorus. As T1 and T2 effects are practically impossible to correct, they may be viewed and exploited as a means of creating "spectral" or "peak" contrast, exactly the same way as in T1 and T2 contrast in MR imaging. Signals in MR images are not acquired under fully relaxed conditions and there are signal losses from T1 saturation and T2 dephasing. The MR images are not corrected for water and fat T1 and T2 effects; their diagnostic value is based on relative changes in image contrast (which depend on T1 and T2) between "normal" and pathological

---

[f]Refer to Section 2.6.1.2 for a discussion of MR visible metabolites.
[g]Besides its effect on spin dephasing, TE also influences the amount of modulation observed on J-coupled peaks.

TABLE 5.1 Relaxation Times T1 and T2 of Major ¹H Metabolites in Normal Adult Brain[a]

	White Matter		Gray Matter	
Metabolite	T1 (ms)	T2 (ms)	T1 (ms)	T2 (ms)
NAA	1380 ± 50	483 ± 20	1330 ± 40	399 ± 9
PCr + Cr	1300 ± 60	209 ± 5	1320 ± 30	204 ± 2
Cho	1440 ± 40	325 ± 10	1390 ± 10	401 ± 20
mI	1200 ± 10	197 ± 1	1130 ± 30	279 ± 14

[a]Adapted from Kreis et al.[10]

tissues, using the same measurement protocol on all patients. In the same way, MRS can play a valuable diagnostic role based on relative changes in metabolite levels, measured with the same protocol, between patients and "normal volunteers," and on serial examination of the same patient.

Coil loading, hence coil sensitivity, varies significantly from one individual to another. Coil loading must be taken into account when comparing MR signal intensities from different subjects, because such a comparison is meaningful only if the two signals were acquired with the same coil sensitivity. In addition to its effect on the received MR signal, coil loading influences the amplitude of the transmitted rf pulse that is needed to produce a given flip angle. It can be shown that the fully relaxed MR signal intensity is inversely proportional to the amplitude of a reference nonselective 90° rf pulse, and that the product of a measured spectral peak area and the amplitude of this pulse is a constant quantity.[12] This has been employed for normalization of peak areas in a given brain spectrum by multiplying the measured peak areas by the amplitude of the 90° nonselective reference pulse from the corresponding measurement, or a factor proportional to this.

The peak area from a selected volume depends on the excited volume size

TABLE 5.2 Relaxation Times T1 and T2 of Major ¹H Metabolites in Various Regions of the Normal Adult Brain[b]

	Insular		Occipital		Thalamus		Cerebellum	
Metabolite	T1 (ms)	T2(ms)	T1 (ms)	T2(ms)	T1 (ms)	T2(ms)	T1 (ms)	T2(ms)
NAA	1650	330	1450	450	1400	340	1700	300
PCr + Cr	1750	250	1550	240	1750	200	1500	190
Cho	1100	380	1150	330	1200	320	1500	410
Ins	1350	130	900	110	1100	150	1850	130

[b]Adapted from Frahm et. al[11]

and on the slice selection profile. Volume calibration can be performed using a phantom solution of known concentration. Following the patient MRS examination, the phantom is placed in the magnet and a signal is measured from the same volume size and location used in the preceding patient exam. Alternatively, in comparative studies, the errors from volume selection are minimized by using the same measurement protocol on all patients, including volume size and localization technique.

Other factors that affect the MR signal and hence the peak area calculation are inhomogeneities in the B_0 and B_1 fields. Effects of B_0 inhomogeneities are minimized by shimming and achieving the best homogeneity possible prior to each MRS measurement. The spatial variation in B_1 (or flip angle) within a volume coil is derived from the variation in signal intensity of the same size volume placed in different locations. Variations of the order of 5% or less have been measured in standard head coils.[19,21] The effects of B_1 inhomogeneities are commonly neglected when reporting clinical MRS data from volume coils. Correction for B_1 inhomogeneities with surface coil measurements can be applied using B_1 maps. This correction is not needed when the rf excitation with surface coils is performed with B_1 insensitive adiabatic pulses.

An important factor in the calculation of metabolite concentrations comes from partial volume effects that are impossible to eliminate in clinical applications. While methods[12,13] that correct for these effects exist, they require additional measurements and are currently impractical for clinical applications. Partial volume effects in localized MRS are minimized by selecting appropriate volumes of interest that avoid a mixture of tissues.

Experience is showing that correction factors may be minimized by well-maintained and adjusted hardware, and a clever and careful design of measurement protocols including consistency and reproducibility checks. Clinically useful MRS data interpretation of relative metabolite levels may be applied without performing all the corrections that are needed for determination of absolute concentrations. Semiquantitative spectral analysis based on peak areas normalized for coil loading and the excited VOI is providing clinically valuable diagnostic tools. Normalized peak areas allow comparison of relative metabolite levels between patients and in serial examinations of the same patient. For example, in a recent ^{1}H MRS study aimed at differentiating recurrent or residual brain tumor from delayed cerebral necrosis in children, a diagnostic index is found from the normalized peak areas of choline and creatine compounds.[15]

5.3.3.3. Peak Area Ratios

An easy assessment of variations in metabolite levels consists of evaluating metabolite peak area ratios. These are calculated from a spectrum of abnormal tissue, and then compared to the corresponding ratios from normal tissue in the same organ and patient, if available. Otherwise, they may be compared to normative values from healthy volunteers. When comparing peak areas of spectra from different individuals and from different measurements, it is important to correct for factors that affect area calculations (eq. 5.4). Some of these factors, such as coil loading, cancel out when ratios of metabolite peak areas from the same spectrum are calculated.

The peak area ratio method is currently used in clinical MR spectroscopy mostly because it is simple and requires no technical expertise or software besides that supplied with the imager. The disadvantage of using ratios is that changes in either or both metabolite concentrations affect the ratio. In addition, both metabolites may vary in a way that leaves the ratio unchanged with respect to a "normal" value. In order to improve the precision and accuracy of this technique, one of the measured metabolites must fulfill the requirement of a reference that does not vary with physical condition and disease. For example, peak area ratios in ^{1}H MRS spectra of the brain have been reported with respect to the Cr/PCr peak based on its presumed relative stability compared to other metabolites. This assumption must be reconsidered since investigations have proven it to be incorrect.[16,17] The second internal peak standard that has been proposed is the unsuppressed water peak, since this can be obtained in a single acquisition. In reality, a stable internal reference may not exist for some organs and conditions. Nonetheless, knowledge of biochemical processes and familiarity with MRS techniques and parameters that affect the spectral peaks have contributed to the successful utilization of ratios in the interpretation of clinical MRS data. In a study investigating the prediction of outcome in acute central nervous system injury, it was demonstrated that clinical examination combined with ^{1}H MRS results, based on measured peak area ratios, classified up to 100% of the patients in their correct outcome groups.[18,19] Semiquantitative methods will remain a valuable alternative for the evaluation of clinical MRS data even after standardized and reliable quantitation techniques become available. Furthermore, there is a reasonable agreement in the literature among identical MRS studies reported as peak area ratios, a fact that will encourage and expand the clinical application of MRS.

5.3.4. Absolute Quantitation

For years, high-field MR spectroscopy has been a valuable quantitative analytical tool in the chemical and biochemical sciences. High-field spectra of solutions are taken under conditions of high B_0 homogeneity, excellent peak separation, and acquisition times limited only by sample or machine stability. Despite the significant progress made in the last few years and its common use at various research institutions, the absolute quantitation of noisy, poorly separated peaks in clinical low-field spectra poses much more formidable problems. It is not surprising that absolute quantitation has yet to reach the stage of routine clinical applicability.

Absolute quantitation is easy to implement reliably on a high-resolution laboratory spectrometer. Sample peak areas can be calibrated against a standard of known size and composition. The standard and the sample of interest are always subjected to the same measurement conditions, including field homogeneity, positioning within the coil, and within the magnet. The relaxation effects of T1 and T2 on the spectra can be eliminated or accurate corrections are applied. These conditions allow direct comparison of the spectral peak areas and calculation of the unknown concentration of the sample from the known concentration of the standard.

Because of the practical difficulties of measuring T1 and T2 on every patient and for every VOI, the high-field quantitation method described above is very difficult to implement on a clinical scanner. Another difficulty encountered on clinical MR units is that measurement conditions vary between the standard sample and a patient and from one individual to the next. One dominant variable is rf coil loading, which was described earlier. The rf coil loading, and hence its sensitivity, varies with the object inserted in the coil. Peak areas have to be corrected for this effect. Alternatively, data from the patient and the standard are measured under the same coil loading condition while both are positioned in the coil. Simple implementation of both methods have been described in the literature.[12,20,21] Table 5.3 illustrates the effect of accounting for correction factors in the calculation of absolute metabolite concentrations.

In vivo absolute quantitation of metabolite amounts uses two types of standards: an internal reference that consists of a metabolite that already

TABLE 5.3. Effect of Correction Factors on the Calculation of the Absolute Concentrations of Major ^{1}H Metabolites in the Parietal Region of Normal Adult Brain[a]

Metabolite	Not corrected (mmol kg^{-1} wet weight)		Corrected[b] (mmol L^{-1})	
	White Matter	Gray Matter	White Matter	Gray Matter
NAA	7.8 ± 0.7	9.8 ± 1.4	8.8 ± 1.0	11.7 ± 2.2
PCr + Cr	5.3 ± 0.7	6.6 ± 1	6.1 ± 0.8	8.2 ± 1.4
Cho	1.6 ± 0.3	1.1 ± 0.3	1.8 ± 0.3	1.4 ± 0.3
mI	3.9 ± 0.9	4.7 ± 0.8	4.7 ± 1.0	6.2 ± 1.1

[a]Adapted from Michaelis. [11]Proton MRS measurement were made with a STEAM sequence (TR = 6000 ms and TE = 20 ms).
[b]Spectral resonances peak areas corrected for differences in coil loading, T2 attenuation, and partial volume effects.

exists in the measured tissue, or an external reference placed in a sealed vial and mounted on the rf coil. The most commonly used internal reference in ^{1}H MRS[22,23] is the water content of tissue, which has also been employed in ^{31}P MRS. With ^{31}P MRS, the differences in measurement conditions between the standard and the sample must be accounted for. Those differences include the resonant frequency, flip angle, receiver sensitivity, and coil loading. With ^{1}H MRS, implementation of the method is simpler since no frequency related corrections are needed. A water-suppressed spectrum is acquired from a selected voxel. Another fully relaxed spectrum with no water suppression is acquired from the same voxel and with the same measurement parameters. The unknown metabolite concentrations are derived from the known water concentration by comparing the corresponding peak areas. Viability of the internal reference technique depends on an accurate knowledge of tissue water content, a homogeneous water distribution, and a constant water concentration. Any one or all of these requirements may not be met, leading to inaccurate estimation of the absolute metabolite levels.

The measurement procedure for the external standard method is similar to the one described above, except that the reference spectrum is acquired from an external standard solution. It is important to use the same measurement sequence and parameters for both measurements. The external standard method is sensitive to $\mathbf{B_0}$ and $\mathbf{B_1}$ inhomogeneities due to the separation of the standard and the measured volume. Differences in flip angles need to be accounted for when calculating the absolute concentrations. An advantage of this method is that it allows better control over the composition of the standard.

In addition to being clinically impractical, absolute quantitation of in vivo metabolites depends on a number of unknown parameters including parts of the metabolites that may be nondetectable or MR invisible.[h] Therefore, absolute quantitation is difficult to measure accurately. This, however, should not be interpreted to mean that there is no clinical utility for MRS. As discussed above, assessment of relative metabolite levels with semiquantitative techniques is already making a difference in patient management and will continue to be used because of its combined relative simplicity and clinical utility.

5.4. EDDY CURRENTS AND MR SPECTROCSOPY

Rapid switching of magnetic field gradients produce electric currents known as eddy currents in conducting structures surrounding the magnet, mainly in the cryogenic shields that are part of superconducting magnets. If not compensated or corrected for, these currents distort the MR signal creating

[h]Refer to Section 2.6.1.2 for a discussion of MR visible metabolites.

spectral artifacts. Several approaches are used to minimize eddy current effects. Many MR scanners are equipped with active shielded gradients that minimize eddy currents in the cryoshield. Hardware compensation consists of adjusting the gradient pulse profile in a way that cancels out eddy current effects. However, the cancellation may be incomplete and the residual effects must be removed with software routines that estimate signal distortions and correct for them.[25–27]

Software approaches to eddy current correction are based on measuring the distortions of a reference signal. The phase evolution and frequency shift of the reference signal is calculated and subtracted from the time domain signal of the metabolites, thus eliminating distortions from the final spectrum. For in vivo ^{1}H MRS, the reference signal may be either a nonsuppressed water signal or the residual water signal from the actual MRS measurement. With the residual water signal method, the reference and the metabolite signals are obtained simultaneously in one measurement thus saving time and ensuring that the measurement conditions are the same for both signals. Following the phase and frequency correction procedures, the residual water signal may be subtracted from the free induction decay, if its amplitude is problematic. This procedure eliminates the residual water peak and the associated broad wings that distort the spectral baseline. (Fig. 5.18a). Otherwise, baseline correction can be used to remove the baseline distortion from residual water. With the nonsuppressed water method, a nonsuppressed water signal is obtained in a separate acquisition, which adds to the exam time. Both the reference signal and the metabolite signals contain phase distortions due to eddy currents. However, motion between the reference scan and the actual measurement may jeopardize the correction procedure.

5.5. CHEMICAL SHIFT IMAGING POSTPROCESSING

Most multivoxel techniques of dimensions greater than or equal to 2 produce a large number of spectra, making individualized processing impractical. Automatic postprocessing programs are essential for CSI and other multivoxel techniques to be feasible. Such automated programs have been developed and used by several research groups, and are currently available on some commercial clinical scanners.

All the postprocessing steps described in this chapter apply to the individual time and frequency domain signals of the most common multivoxel method implemented on clinical imagers, chemical shift imaging (CSI). Because CSI measures spatial as well as frequency distribution, additional manipulations must be performed on the collected raw data either before or after the time to frequency Fourier transformation that generates the spectra.

Because of the multidimensional nature of CSI, processing can require more than one spatial Fourier transformation. One-dimensional[i] CSI data require two Fourier transformations, one to extract the spatial information, and one to extract the frequency information. Two- and three-dimensional CSI data sets require three and four Fourier transformations, respectively.

As explained in Section 4.5.3.1, a 2D CSI measurement with a 16×16 matrix produces spectra from 256 voxels. However, the anatomy of interest may be covered by a smaller number of voxels. To reduce calculation time, the first step in processing CSI data is to define the VOI by registering the MRS data grid on a MRI image. The VOI may be scrolled in order to match particular voxels of the grid to particular abnormal tissue locations on the image. This is called grid or voxel shifting (Section 4.5.3.1). The FID signals are processed with apodization, time, and spatial Fourier transformation. This processing produces spectra assigned to individual voxels. Ideally, the software automatically computes any frequency and phase corrections required for each voxel. Frequency correction of individual spectra may also be needed because local inhomogeneities in large measurement volumes shift the local frequency relative to other parts of the VOI. Automated software algorithms may then be used to calculate peak parameters such as width, position, amplitude, and area. Quantification with CSI follows the same principles discussed in Section 5.3. Additional uncertainties have to be accounted for, such as signal contamination from neighboring voxels due to the pointspread function (Section 4.5.3), and chemical shift offset errors[13] (Section 5.3.3.2). When CSI is used with a surface coil, spatial intensity variations due to the receiving coil profile must be included in the quantification correction factor.[28]

The results from CSI data processing may be displayed as a spectral map (Figs. 4.17 and 6.21) that consists of the spectra drawn in the phase encoding grid and overlaid on a MRI image. This display allows a quick survey of spectral variations over the VOI. Any one of the peak parameters (width, integral, or amplitude) may be examined for spatial variations with the help of a grid display. Another way of presenting CSI results consists of metabolite maps, also referred to as metabolite images, which are calculated from peak integrals and displayed over an MR image. These resemble Positron emission tomography (PET) images. An NAA metabolite image[j] is the same as a conventional MRI image except that it is calculated from the peak integral and spatial distribution of NAA, instead of water and fat. Metabolite images have lower resolution because of S/N considerations; but they are useful for showing the distribution of a given metabolite and for indicating

---

[i]Refer to Section 4.5.3.1 for definition of CSI dimensions.
[j]Metabolite images are shown in Figures 4.17 and 6.22.

areas of elevated or decreased concentrations. At the present state of the art, brain metabolite images from MRS match or exceed the spatial resolution of PET brain images.

REFERENCES

1. R. J. Ogg, P. B. Kingsley, and J. S. Taylor: The line broadening and unambiguous specification of the Gaussian filter, *J. Magn. Reson., Ser. A*, **117**, 113–114, 1995.

2. J. Frahm, H. Bruhn, M. L. Gyngell, K. D. Merboldt, W. Hänicke, and R. Sauter: Localized high-resolution proton NMR spectroscopy using stimulated echoes: Initial applications to human brain in vivo, *Magn. Reson. Med.* **9**, 79–93, 1989.

3. T. Michaelis, K. D. Merboldt, W. Hänicke, M. L. Gyngell, H. Bruhn, and J. Frahm: On the identification of cerebral metabolites in localized ^{1}H NMR spectra of human brain in vivo, *NMR Biomed.* **4**, 90–98, 1991.

4. I. Marshall, J. Higinbotham, S. Bruce, and A. Friese: Use of Voigt lineshape for quantification of in vivo ^{1}H spectra, *Magn. Reson. Med.* **37**, 651–657, 1997.

5. H. Barkhuijsen, H. R. De Beer, W. M. M. J. Bovee, and D. Van Ormondt: Retrieval of frequencies, amplitudes, damping factors and phases from time domain signals using linear least-squares procedure, *J. Magn. Reson.* **61**, 465–481, 1985.

6. S. J. Nelson and T. R. Brown: A method for automatic quantification of one-dimensional spectra with low signal-to-noise ratio, *J. Magn. Reson.* **75**, 229–243, 1987.

7. S. J. Nelson and T. R. Brown: The accuracy of quantitation from 1D NMR spectra using the PIQABLE algorithm, *J. Magn. Reson.* **84**, 95–109, 1989.

8. G. L. Bretthorst: Bayesian analysis. III. Applications to NMR signal detection, model selection, and parameter estimation, *J. Magn. Reson.* **88**, 571–595, 1990.

9. R. Stoyanova, A. C. Kuesel, and T. R. Brown: Application of principle-component analysis for NMR spectral quantitation, *J. Magn Reson. Ser. A* **115**, 265–269, 1995.

10. R. R. Kreis, T. Ernst, and B. D. Ross: Absolute quantitation of water and metabolites in the human brain. II. Metabolite concentrations, *J. Magn. Reson. Ser. B*, **102**: 9–19, 1993.

11. J. Frahm, H. Bruhn, M. L. Gyngell, K. D. Merboldt, W. Hänicke, and R. Sauter: Localized proton NMR spectroscopy in different regions of the human brain in vivo, relaxation times and concentrations of cerebral metabolites, *Magn. Reson. Med.* **11**, 47–63, 1989.

12. T. Michaelis, K. D. Merboldt, H. Bruhn, W. Hänicke, and J. Frahm: Absolute concentrations of metabolites in the adult human brain in vivo: Quantification of localized proton MR spectra, *Radiology* **187**, 219–227, 1993.

13. T. Ernst, R. Kreis, and B. D. Ross: Absolute quantitation of water and metabolites in the Human Brain. I. Compartments and water, *J. Magn. Reson., Ser. B*, **102**, 1–8, 1993.

14. S. Mackay, F. Ezekiel, V. Di Sclafani, D. J. Meyerhoff, J. Gerson, D. Norman, G. Fein, and M. W. Weiner: Alzheimer disease and subcortical ischemic vascular dementia: Evaluation by combining MR imaging segmentation and ^{1}H MR spectroscopic imaging, *Radiology* **198**, 537–545, 1996.

15. J. S. Taylor, J. W. Langston, W. E. Reddick, P. B. Kingsley, R. J. Ogg, M. H. Pui, et al.: Clinical value of proton magnetic resonance spectroscopy for differentiating recurrent or residual brain tumor from delayed cerebral necrosis, *Int. J. Rad. Oncol. Biol. Phys.* **36(5)**, 1251–1261, 1996.

16. B. L. Miller: A review of chemical issues in ^{1}H NMR spectroscopy: N-Acetyl-L-aspartate, creatine and choline, *NMR Biomed.* **4**, 47–52, 1991.

17. B. Ross and T. Michaelis: Clinical applications of magnetic resonance spectroscopy, *Magn. Reson. Q.* **10**, 191–247, 1994.

18. K. L. Auld, S. Ashwal, B. A. Holshouser, L. G. Tomasi, R. M. Perkin, B. D. Ross, and D. B. Hinshaw: Proton magnetic resonance spectroscopy in children with acute central nervous system injury, *Pediatr. Neurol.* **12**, 323–334, 1995.

19. B. A. Holshouser, S. Ashwal, G. Y. Luh, S. Shu, S. Kahlon, K. L. Auld, L. G. Tomasi, R. M. Perkin, and D. B. Hinshaw: Proton MR spectroscopy after acute central nervous system injury: Outcome prediction in neonates, infants, and children, *Radiology* **202**, 487–496, 1997.

20. P. S. Tofts and S. Wray: A critical assessment of methods of measuring metabolite concentrations by NMR spectroscopy, *NMR Biomed.* **1(1)**, 1–10, 1988.

21. J. Hennig, H. Pfister, T. Ernst, and D. Ott: Direct absolute quantification of metabolites in the human brain with in vivo localized proton spectroscopy, *NMR Biomed.* **5**, 193–199, 1992.

22. P. B. Barker, B. J. Soher, S. J. Blackband, J. C. Chatham, V. P. Mathews, and R. N. Bryan: Quantification of proton NMR spectra of the human brain using tissue water as an internal concentration reference, *NMR Biomed.* **6**, 89–94, 1993.

23. P. Christiansen, O. Henriksen, M. Stubgaard, P. Gideon, and H. B. W. Larson: In vivo quantification of brain metabolites by ^{1}H MRS using water as an internal standard, *Magn. Reson. Imag.* **11**, 107–118, 1993.

24. K. R. Thulborn and J. J. H. Ackerman: Absolute molar concentrations by NMR in inhomogeneous B1. A scheme for analysis of in vivo metabolites. *J. Magn. Reson.* **55**, 357–371, 1983.

25. R. J. Ordidge and I. D. Cresshull: The correction of transient B_0 field shifts following the application of pulsed gradients by phase correction in the time domain, *J. Magn. Reson.* **69**, 151–155, 1986.

26. U. Klose: In vivo proton spectroscopy in presence of eddy currents, *Magn. Reson. Med.* **14**, 26–30, 1990.

27. W. R. Riddle, S. J. Gibbs, and M. R. Willcott: Removing effects of eddy currents in proton MR spectroscopy, *Med. Phys.* **19(2)**, 501–509, 1992.

28. S. E. Moyher, B. S. Daniel, B. Vigneron, and S. J. Nelson: Surface coil MR imaging of the human brain with an analytic reception profile correction. *J. Magn. Reson. Imag.* **5**, 139–144, 1995.

CHAPTER 6

Clinical Applications

The integration of MRS into patient care requires expertise in two areas: the successful performance of an MRS patient examination and the accurate interpretation of patient MR spectra. The former requires familiarity with technical issues that affect the collection and processing of spectral data; the latter requires knowledge of metabolites observed in normal and abnormal tissues. Numerous reviews and surveys[1–10] have analyzed the role of MRS in the assessment of pathologies and in therapy monitoring. The reader is referred to those references and to the ones cited therein for information that correlates spectral features to various disease states. In keeping with the concept of the book, this chapter deals with technical issues that affect the outcome of an MRS patient examination. A description of the major metabolite peaks observed from selected normal tissues in vivo is included with representative illustrations of variations seen in spectra from abnormal tissues.

6.1. GENERAL CONCEPTS

Over the years, in vivo MR spectroscopy investigations have examined a variety of nuclei (^{1}H, ^{31}P, ^{19}F, ^{13}C, ^{14}N, and ^{23}Na) and organs with a variety of magnetic field strengths (1.5, 2, 4, and 7 T). This discussion will focus on clinical applications that are feasible today on a standard clinical 1.5 T scanner. Currently, proton magnetic resonance spectroscopy (^{1}H MRS) in the brain is by far the most commonly performed MRS patient examination. It can be performed using standard MR imaging hardware, and the necessary software is readily available on most systems. The goal of ^{1}H MRS is to detect signals from small concentrations of metabolites in the presence of a large concentration of water contained in a small volume over a narrow frequency range. It is critical to have practical and precise localization techniques, the best field homogeneity possible, and effective water suppression. These requirements are met with careful hardware adjustments and with optimization of measurement techniques and parameters. Recently introduced automated software procedures have considerably simplified the adjustment steps and shortened the MRS examination time. Most commercial MRS software includes some version of proton Single Voxel Spectroscopy (SVS) and volume-selective Chemical Shift Imaging (CSI). Spectroscopy with phosphorus or other nuclei requires special hardware that may be added to the standard configuration of the scanner at the time of initial installation or at a later time. This hardware consists of the appropriate coils and the related circuitry for rf pulse transmission and signal reception at the specific frequencies. Phosphorus localization software such as slice selective CSI and ISIS is commercially available on some scanners. Software features and details depend on a particular system and manufacturer.

6.2. CLINICAL MRS: GETTING STARTED

When starting a clinical MRS program, it is important to accumulate MRS data on 15–20 volunteers using the measurement protocol that is intended to be used in patient examinations. This procedure has two purposes. The first is to build up experience and confidence in operating the MRS software and the second is to establish a baseline of normal volunteer data that will serve as a reference in the interpretation of patient data. This information also constitutes a standard of hardware operation in addition to the regular quality assurance (QA) measurements that should be performed on the scanner regularly using a phantom solution. Simple and fast daily QA measurements, similar to those performed for imaging purposes, ensure reproducibility and help detect changes in hardware performance that may affect MRS results. The measurement can be as simple as acquiring an FID signal from the same QA phantom and measuring its amplitude. At later times, when a measurement protocol is changed, 10 volunteer measurements may be sufficient to establish the needed baseline. In addition, it is advisable to implement simpler techniques first, such as ^{1}H MRS single voxel measurements in diffuse brain disease with long TE, and then with short TE sequences. With the accumulated experience, examinations of focal diseases with SVS or CSI techniques will become easier and faster to implement.

6.3. HYDROGEN SPECTROSCOPY

6.3.1. Brain

As in imaging, there are technical reasons that make the brain easier than other organs to examine with MRS: motion artifacts are minor; shimming is relatively easy to perform, especially with currently available automated software; and there are no detectable (i.e., mobile) lipids in normal brain tissue itself. Mobile lipids may appear as part of pathology such as necrotic tumors. Another advantage of ¹H MRS in the brain is that it does not require a special coil and may be incorporated at the end of a standard brain MRI.

6.3.2. ¹H MRS Brain Exam: Measurement Steps

While MRS examinations have been simplified by the development of auto-mated software, there are advantages to understanding the concepts underly-ing the various steps of the adjustment and measurement processes. With such information, an operator can identify possible causes of error and take corrective actions, or assume manual control of the examination when auto-mation fails. The result is an increased number of successful MRS exams.

The patient screening procedures that are commonly used prior to an MRI examination are also appropriate for MRS. The first three steps listed below are not needed if the MRS examination is an extension of an ongoing MR imaging scan. The sequential steps of an MRS brain examination are:

- Patient positioning.
- Global shimming (or alternative approaches).
- Acquisition of MR images for localization.
- Selection of the MRS measurement sequence and parameters.
- Selection of a volume of interest.
- Localized shimming.
- Optimization of water suppression.
- MRS data collection.
- MRS data processing and display.

6.3.2.1. Patient Positioning

The patient is positioned in the standard head coil, in the same way as for a brain imaging exam (supine, head first). The head should be firmly sup-ported and secured in a way that prevents involuntary motion, which can result in line broadening, incorrect localization, or loss of signal. In SVS, loss of signal is due to collecting spectral data from outside the excited volume. In phase-encoded techniques such as CSI, motion results in extra-neous phase differences, the type which produces ghosting on MR images. With techniques that rely on combining data from multiple measurements such as ISIS, motion between measurements leads to incomplete cancella-tion of unwanted signals.

6.3.2.2. Global Shim

Global shimming is a procedure for optimizing the magnetic field homogeneity over the entire volume detected by the receiver coil. Automated shimming procedures are currently available and routinely used at the beginning of an imaging or spectroscopy patient examination. These procedures are fast and ensure good quality images and spectra. In MRS, global shimming provides the starting shim values for localized shimming. Coil tuning, transmitter frequency and power adjustments are normally performed prior to starting the shim procedure. Because shimming modifies the external magnetic field B_0, a frequency adjustment is usually needed after a satisfactory shim is obtained.

An alternative approach to performing a global shim, which may be used to speed up patient MRS examinations, is to store shim sets as data files that may be recalled and used when needed. The files are obtained from shimming on volunteers having different head sizes. For example, a "small head" shim file may be created, saved, and called up for use on people having small size heads such as children or small adults. Until recently, this was a common practice because only manual shimming, which may be a tedious and lengthy procedure, was available on clinical MR scanners. With the advent of imaging techniques that require optimal field homogeneity, such as fat saturation and echo planar imaging (EPI), automated global shimming is often incorporated into the routine adjustment procedures that are performed at the start of each MR imaging examination. There is no need to repeat the global shim if the MRS measurement is appended to such an imaging study. However, it is still a good practice to store shim files from volunteers, and to use them on patients when the automated shim procedure fails to produce the desired homogeneity.

6.3.2.3. Scout Images

Scout views are used to guide the clinician or spectroscopist in the selection of a volume of interest. The ability to display the localization region and position the VOI on MR images is part of commercial clinical MRS software (Fig. 6.1). When available, images from preceding MR imaging studies in the examination may be used for localization.

6.3.2.4. Selection of the Measurement Technique and Parameters

The higher spatial resolution of CSI compared to SVS and its ability to examine regional metabolite variations from a single measurement make it the method of choice for all pathologies. Nonetheless, SVS is more commonly used in clinical routine practice because of its simplicity, widespread availability, and proven clinical utility. Choice of the measurement technique

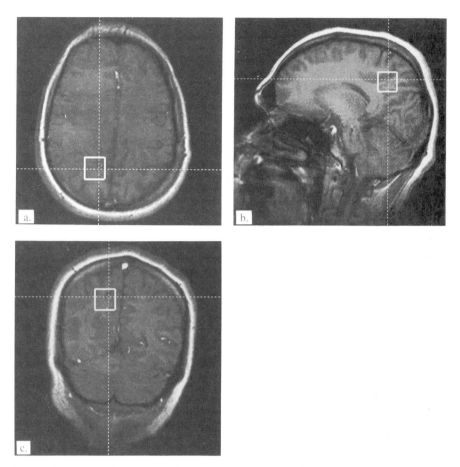

Figure 6.1. Positioning of a voxel on transverse (*a*), sagittal (*b*), and coronal (*c*) images.

depends on the pathology to be examined and on the availability of a given technique on a particular scanner. In general, SVS is employed with diffuse global disease or with single focal lesions. Chemical shift imaging is more appropriate for examining pathology that is multifocal or requires high spatial resolution to distinguish pathology accurately. A single CSI measurement produces spectra from the lesion and from normal tissue. In addition, CSI offers better resolution and allows grid shifting[a] during data processing. However, if CSI is not available, a SVS exam may be performed on a focal lesion by acquiring spectra from the lesion and from the contralateral side. Some heterogeneous lesions, such as large tumors, benefit from examination of more than one voxel, in order to distinguish the more active from the more necrotic regions.

[a]Refer to Sections 4.5.3.1 and 5.5 for a description of grid shifting.

TABLE 6.1. Effect of Measurement Parameters on S/N and Resolution[a]

Parameter	Variation	S/N[b]	Resolution
TE	↑	↓	
TR	↑	↑	
No. of acquisitions	↑	↑	
Voxel size	↑	↑	↓ Spatial resolution
Vector size	↑		↑ Spectral resolution

[a]The S/N variations with measurement parameters of MRS PRESS and STEAM sequences are the same as for any imaging sequence. The parameter "vector size," which determines the number of sample points of the signal, is equivalent to the number of read out points in an imaging sequence.
[b]For the same parameters, a PRESS sequence has twice the S/N as a STEAM sequence.

As in imaging, the choice of measurement parameters in MRS has a direct influence on the results of the examination (Table 6.1). The parameters that may be selected are the echo time (TE), the repetition time (TR), the number of acquisitions, the voxel size and position, and the number of data points used to sample the time domain signal. The echo time, repetition time, number of acquisitions, and voxel size all affect S/N. The number of sample points, referred to as vector size, determines the spectral resolution and is typically set to 1024 or 2048. The most common TE values for ¹H MRS are 20 or 30 ms for a STEAM sequence, and 135 or 270 ms for either STEAM or PRESS.[b] The PRESS sequences with shorter TEs are also being implemented on scanners with shielded gradient systems. Long TE spectra (Fig. 6.2) have several advantages: peaks from the major brain metabolites (NAA, choline, or creatine) are visible as well from lactate, if present; peaks are suppressed from lipids and from *J*-coupled protons such as those from glutamate, glutamine, γ-aminobutyric acid (GABA), and inositols; and spectra are simple to process accurately for peak areas. Shorter TE sequences (Fig. 6.3) yield additional information from metabolites with short T2 whose variations may be critical to certain diseases such as hepatic encephalopathy.[11,12] Short TE sequences are being employed more often because of the added information. When only variations in the major metabolite peaks (NAA, Cho, or Cr/PCr) need to be determined, or when only lactate is being investigated, longer TE sequences offer the benefit of a simplified spectrum with reduced contribution from lipid signals at the frequency of the lactate doublet. The absence of peak interference facilitates interpretation and quan-

[b]Both STEAM and PRESS are described in Chapter 4, along with the advantages and disadvantages of each sequence.

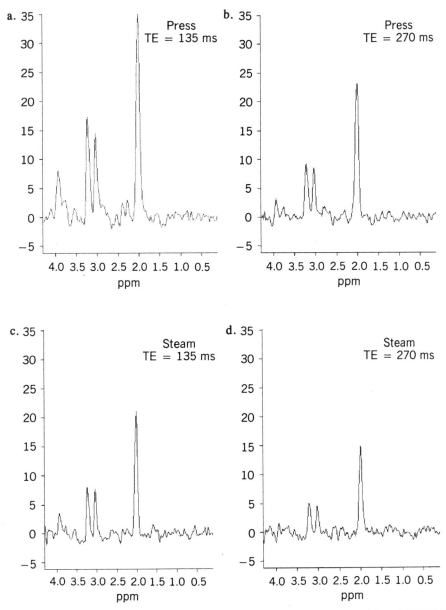

Figure 6.2. Normal brain spectra from an 8-cm³ voxel acquired with PRESS and STEAM sequences using a TR of 1500 ms and 256 acquisitions. The same receiver gain and the same processing procedure were used on all spectra. For the same TE, PRESS has more S/N than STEAM. Longer TE for the same sequence produces smaller S/N.(*a*) PRESS with TE = 135 ms. (*b*) PRESS with TE = 270 ms. (*c*) STEAM with TE = 135 ms. (*d*) STEAM with TE = 270 ms.

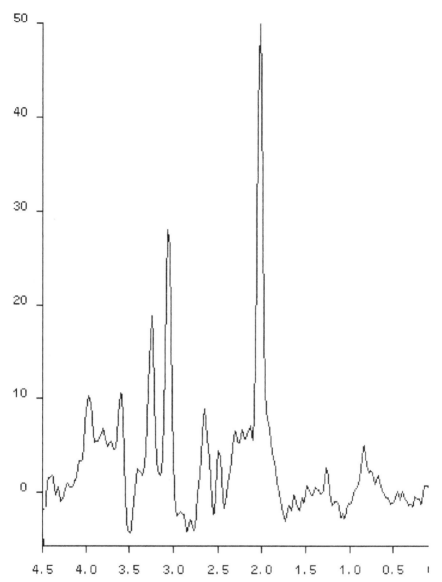

Figure 6.3. A normal brain spectrum from an 8-cm³ voxel acquired with a short TE (20 ms) STEAM sequence and a TR of 1500 ms. A detailed description of the peaks is found in the text.

TABLE 6.2. Typical Measurement Parameters for ¹H MRS in the Brain

Parameter	Typical Values Long TE Measurement	Typical Values Short TE Measurement
TE (SVS or 2D CSI)	135 or 270 ms	20 or 30 ms
TR (SVS or 2D CSI)	1500–3000 ms	1500–3000 ms
Vector size (SVS or 2D CSI)	1024–2048	1024–2048
Voxel size (SVS)[a]	2–27 cm³	2–27 cm³
No. of acquisitions (SVS)	128–256	64–256
Measurement time (SVS)	~ 3 to ~ 6 min	~ 1.5 to ~ 6 min
Voxel (2D CSI)[b]	1–2 cm³	1–2 cm³
No. of acquisitions (2D CSI)	1 or 2	1 or 2
Measurement time (2D CSI)	~ 6 to ~ 12 min	~ 6 to ~ 12 min

[a]For smaller voxel size, it is preferable to use CSI techniques because of rf profile and S/N considerations.
[b]The 1 cm³ resolution is obtained on standard volume head coils. Smaller voxels may be obtained with ¹H CSI using surface coils.

titation of the spectrum. Choice of the two long TE values[c] (135 and 270 ms) is based on the fact that the lactate doublet is inverted at 135 ms and upright at 270 ms with respect to other peaks.[d] It is to be noted that the amplitude of the lactate doublet is 40% smaller at TE = 135 ms than at TE = 270 ms.[13] This difference is due to the fact that refocusing of the antiphase magnetization at 135 ms is not as efficient as refocusing of the in-phase magnetization at 270 ms. Therefore, improved sensitivity to lactate detection is achieved with a PRESS sequence (which has twice the S/N as STEAM) and with a TE of 270 ms.

Good S/N spectra are obtained with a TR of 1500–3000 ms and with 128–256 acquisitions (Table 6.2) depending on TE, the sequence type and the voxel size. Smaller voxels and shorter TRs require more acquisitions. The effects of variations in some of the measurement parameters are illustrated in Figs. 6.4–6.6. As in imaging, the S/N is directly proportional to the voxel size and to the square root of the number of acquisitions. Thus, reducing the volume size by a factor of 2 requires an increase in the number of acquisitions by a factor of 4^2 in order to keep the same S/N; the scan time is also increased by a factor of 4. Use of the same sequence and the same set of parameters on patients with similar pathologies results in more efficient data interpretation and comparison.

[c]As indicated in Section 2.6.1.2, an accurate analysis of the lactate *J*-coupling yields TE values of 144 and 288 ms instead of the 135 and 270 ms that have been widely used in clinical practice and in the literature. The difference is significant for implementation of lactate editing techniques, but does not have a noticeable effect on other metabolite peaks.
[d]Refer to figures 6.21 and 6.15 for illustrations of the lactate doublet at TE = 135 ms and TE = 270 ms, respectively.

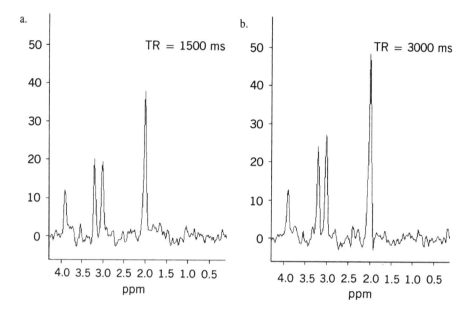

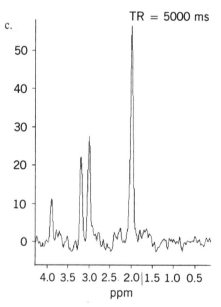

Figure 6.4. Effect of TR on spectrum. Note the improved S/N at longer TR. Spectra are acquired from an 8-cm³ voxel with PRESS using a TE of 135 ms and 256 acquisitions. The same receiver gain and the same processing procedure were used on all spectra. (*a*) TR = 1500 ms. (*b*) TR = 3000 ms. (*c*) TR = 5000 ms.

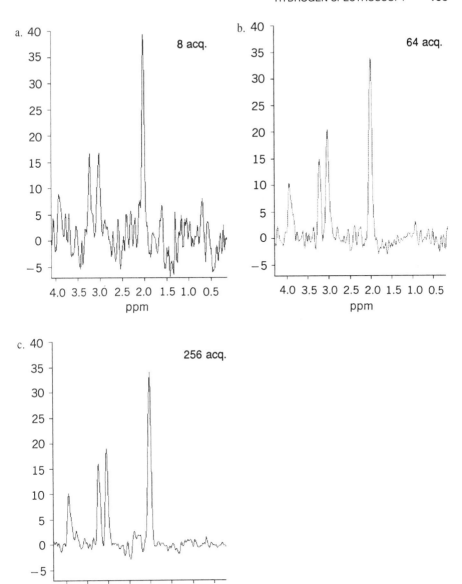

Figure 6.5. Effect of the number of acquisitions on spectrum. Note improved S/N with more acquisitions. Spectra are acquired with a PRESS sequence at a TR = 1500 ms and a TE = 135 ms, on an 8-cm³ voxel and with the same receiver gain. The same processing procedure was used on all spectra. (*a*) 8 acquisitions. (*b*) 64 acquisitions. (*c*) 256 acquisitions.

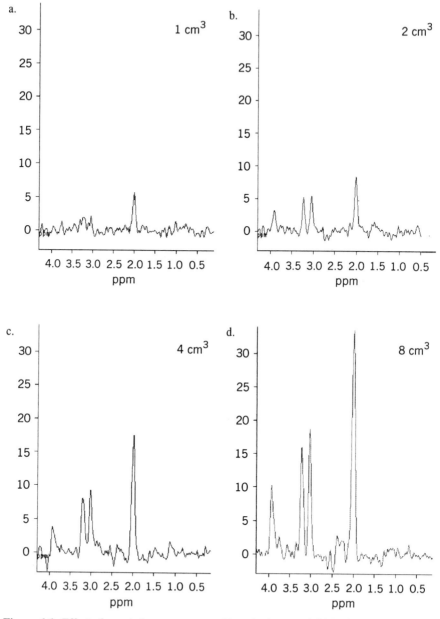

Figure 6.6. Effect of voxel size on spectrum. Note the improved S/N with increased voxel size. Spectra are acquired on the same volunteer with a PRESS sequence at a TE = 135 ms, TR = 1500 ms, and with 256 acquisitions. The same receiver gain value was used on all measurements. (*a*) 1-cm³ voxel. (*b*) 2-cm³ voxel. (*c*) 4-cm³ voxel. (*d*) 8-cm³ voxel.

TABLE 6.3. Detectability of ^{1}H Normal Brain Metabolites[a]

Metabolite	Peak Position (ppm)	TE	Visible in Normal Brain
Lipids	0–2	Short	As contamination
Lactate	1.33	Short and long	Not detectable
Alanine	1.48	Short and long	Not detectable
NAA	2.01, 2.60	Short and long	Yes
Glx	2–2.4, 3.6–3.8	Short	Very small
Cr/PCr	3.02, 3.94	Short and long	Yes
Cho	3.22	Short and long	Yes
Glucose	3.43, 3.80	Short	Hardly detectable
mI	3.56, 4.06	Short	Yes

[a]Chemical shifts are from Michaelis et al.[18], with water defined at 4.7 ppm.

Volume selective 2D chemical shift imaging uses the same TR, TE, and number of data points as single voxel measurements. As in imaging, the slice thickness, field of view, and matrix size determine the spatial resolution and the S/N ratio. Typical parameters for a 2D CSI brain scan are a slice thickness of 1–2 cm, a field of view (FOV) of 16–24 cm, and a matrix of 16 × 16. Therefore, the in-plane resolution is 1–1.5 cm in each direction, and the individual voxels are 1–4.5 cm^3 depending on the FOV and slice thickness. A phase encoding matrix size of 16 × 16 and a TR of 1500 ms result in a scan time of 6.5 min. A 32 × 32 matrix yields higher resolution at the expense of a longer scan time of about 26 min.

6.3.2.5. Position of the VOI

Positioning of the VOI is dictated by the pathology to be examined and the clinical questions to be answered. When examining a focal disease with SVS, the voxel is placed over the abnormal tissue as seen on the MR images. The size of the voxel is adjusted in order to minimize inclusion of normal brain tissue and reduce partial volume averaging. Avoiding inclusion of normal tissue within an 8 cm^3 rectangular voxel is difficult when examining small or irregular lesions. The smaller resolution and flexibility in data processing make CSI better suited for examination of irregularly shaped regions of pathology and focal diseases in general. Placement of a single voxel is easier when assessing diffuse brain disease or a single large focus. For diffuse encephalopathies, a VOI (8–27 cm^3) may be selected in white or in gray matter, depending on which one may be affected most by the disease in question. In some cases, it is beneficial to examine both white and gray matter voxels. As reported by several publications,[14,15] there are modest differences in metabolite concentrations of white and gray matter (Section 6.3.3). Although it is difficult to completely avoid partial volume effects, minimizing the mixture of gray and white matter and excluding CSF simplify interpretation of the spectrum. Careful positioning away from the scalp reduces spectral contamination from unwanted lipid signals. Avoiding bones, sinuses, and blood vessels, if possible, allows for a good shim and optimum spectral quality. It is best to consistently measure the same voxel position in all patients suffering from the same diffuse global disease. This facilitates comparison of spectra and allows reproducibility checks.

When examining a focal disease with SVS, one or more voxels should be measured from the lesion and one voxel from the contralateral side. With volume selective 2D CSI, a slice 10–20 mm thick, usually transverse, is positioned over the lesion on a sagittal or coronal image; a VOI is then defined in the plane of the slice. The VOI should be large enough to include both the lesion and some normal tissue, but small enough to be completely enclosed within the brain, away from the scalp. For 2D CSI sequences relying on outer volume saturation to suppress scalp lipid, careful positioning of the four pairs of saturation pulses is essential. It is easier to obtain a good shim and better spectra on a smaller VOI. Figure 6.12(*a*) illustrates positioning of two small VOIs for the examination of the temporal lobe regions of epileptic patients. This scheme was designed[16] in order to optimize the local shim and ensure good spectral quality, which is more difficult to achieve with a single large VOI covering both temporal lobe regions due to susceptibility effects from sinuses and tissue heterogeneity. Another example of positioning a CSI volume for examination of a focal lesion is illustrated is Figure 6.20.

6.3.2.6. Local Shim

The purpose of local shimming is to optimize the magnetic field homogeneity specifically over the selected VOI. A good local shim produces narrower metabolite peaks, better spectral resolution, and improved S/N. Local shimming is performed with the selected MRS sequence and uses the water signal in the absence of any suppression. Starting from the global shim values, the procedure consists of varying the shim currents, one at a time, until the narrowest water peak is produced. The full width at half-height of the water peak is used as a measure of the shim quality. A local shim of 4–10 Hz may be achieved on voxels of 8–27 cm^3, depending on the size and position of the voxel. Areas of different magnetic field susceptibilities should be avoided due to induced line broadening.

6.3.2.7. Water Suppression

In previous steps, the water signal has been used for shimming and for adjustment of system parameters. However, the actual collection of metabolite data requires that water suppression be employed. A general discussion of various approaches to water suppression in ^{1}H MRS is presented in

Section 4.6. The standard water suppression methods implemented on commercial scanners employ CHESS pulses that typically have a 60-Hz bandwidth. Optimization of water suppression is achieved by varying the amplitude of the CHESS pulses to minimize the water peak height. Alternatively, since only a 50–100-fold reduction of the water peak amplitude is needed, incomplete suppression may be used. The residual water signal that remains is employed during data processing for frequency and phase correction prior to its subtraction from the final spectrum (Fig. 5.4).

6.3.2.8. Measurement

The MRS measurement is started with the optimized local shim, optimized water suppression pulses, and a maximum receiver gain. The measurement time varies from 3 to 6 min for SVS and up to 12 min for CSI, depending on the measurement parameters. The resulting MRS data consist of the real and imaginary parts (Fig. 5.7) of the free induction decay or time domain data.

6.3.2.9. Data Processing and Display of the Spectrum

The MRS data is processed either automatically or manually following the steps described in Chapter 5. The object of data processing is to convert the time domain signal into a frequency domain signal, extract the absorption spectrum, and interpret and quantify the peaks. The spectrum is typically displayed with the water peak set at 4.7 ppm, which is the chemical shift of water with respect to tetramethylsilane (TMS), the accepted chemical shift standard for ^{1}H MR spectroscopy. In this convention, the major metabolite peaks occur between 0 and 4.3 ppm, which is the range commonly used for displaying ^{1}H MRS brain spectra.

6.3.3. ^{1}H MRS Brain Spectra

Figures 6.2 and 6.3 show typical ^{1}H MRS spectra of the normal human brain from a clinical 1.5-T MR scanner. The long TE spectra (TE = 135 and 270 ms) are characterized by three major peaks:[17,18] NAA at 2.02 ppm, creatine/phosphocreatine at 3.0 ppm, and choline at 3.2 ppm. N-Acetyl aspartate (NAA) is the dominant peak in normal adult brain spectra. It is accepted as a neuronal and axonal marker whose physiological role is currently unknown. It has been shown to increase during brain development after birth and in childhood and to decrease[5,19–22] in old age. Reduced NAA has been observed with many neurological (Table 6.8) diseases that cause neuronal and axonal degeneration: epilepsy, dementias, stroke, hypoxia, multiple sclerosis, and many leukoencephalopathies. Increased NAA is only found in children with Canavan's disease.[23–25] This genetic disorder results in a deficit in the enzyme that breaks down NAA. Evidence has been found suggesting that the behavior of this peak may be affected by other metabolite peaks occurring at the same location. The peak at 2.02 contains a small contribution from N-acetyl-aspartate-glutamate (NAAG)[18,26] and other metabolites; it is sometimes referred to as the N-acetyl group peak instead of simply NAA. In addition to the dominant peak at 2.02 from the methyl (CH$_3$) group,[e] a smaller NAA peak from the methylene (CH$_2$) group can be seen at 2.6 ppm in normal brain spectra of high S/N.

The major peak at 3.02 is from the CH$_3$ group of creatine and phosphocreatine (Cr + PCr), referred to as total Cr. Another Cr + PCr peak, from the CH$_2$ group, can be observed at 3.94 ppm if good suppression of the water peak at 4.7 ppm is achieved; however, only the peak at 3.02 is typically employed in the interpretation of spectral data since it has a larger area. Total Cr has been considered to be stable enough to be used as an internal reference in reporting relative concentrations of other brain metabolites, but recent findings suggest that this assumption should be used with care.[4] Examples of total Cr variations include an increase with trauma and a decrease with hypoxia,[27] stroke,[28–30] and tumor.[31]

The peak at 3.22 is a mixture of choline and choline-containing compounds. Choline is thought of as a product of myelin breakdown. The peak is generated by the nine protons in the (CH$_3$)$_3$ group of the choline molecule and contains information related to cell density. Choline is the dominant peak in long TE spectra of normal neonate brain (Fig. 6.7a). In adult brain, an increase in the Cho peak area is associated with Alzheimer's disease,[32–34] chronic hypoxia,[4] postliver transplant,[4] and epilepsy,[16,35] while a decrease is seen in hepatic encephalopathy.[11,12] Malignant glial tumors and a variety of primary brain tumors also show elevated Cho resonances.[31] Several questions remain to be answered regarding the choline peak composition and its variation with metabolic processes.

Lactate, when present, appears at 1.33 ppm as an inverted doublet with PRESS and a TE = 135 ms. The doublet is upright with PRESS at TE = 270 ms and with STEAM at any TE. Having a concentration of about 1 mM, lactate is usually not seen in the spectra of normal brain. The signal is from protons of the CH$_3$ group. Lactate has been detected in patients with stroke,[36–38] some brain tumors, hypoxia, anoxia, and mitochondrial encephalopathies.[39–41] It has been observed to increase in the epileptic focus immediately following a seizure.[42] Lactate production has been measured with ^{1}H MRS during the early stage of focal brain activation.[43] With physiological photic stimulation a transient increase was reported in lactate levels of the visual cortex.[44] Alanine is another resonance that is not usually detected in normal brain; however, it has been observed in some meningiomas.[45,46] This signal is a doublet from protons in the CH$_3$ group and occurs at 1.48 ppm.

---

[e]Chemical structures representing the various brain metabolites are found in Appendix B.

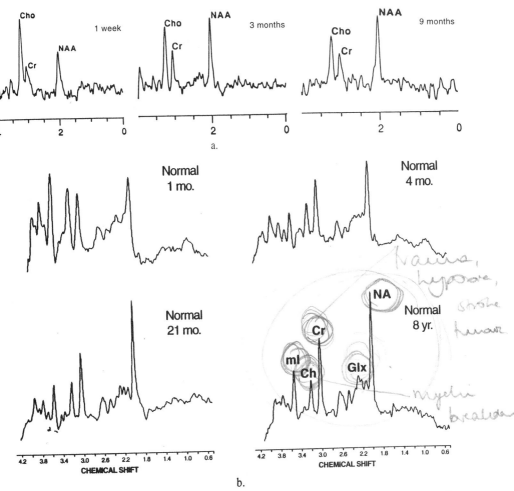

Figure 6.7. (*a*) Proton MR spectra from the brain in normal development at age one week (left spectrum), 3 months (middle spectrum), and 9 months (right spectrum). The spectra were obtained with a STEAM sequence (TR = 2000 ms, TE = 272 ms). (Reproduced by permission of RSNA Publications from Kimura et al., *Radiology 194*, 483–489, 1995). (*b*) Proton MR spectra (3000/20 ms, STEAM) obtained in the occipital gray matter in four developmentally normal patients. Spectral peaks are assigned to NAA (NA) (2.0 ppm), glutamate and glutamine (Gl_x 2.10–2.45), Cr (3.0 ppm), Cho (3.2 ppm), and myoinositol (mI, 3.5 ppm). (Reproduced by permission of RSNA Publications from B. A. Holshouser, et al., *Radiology, 202*, 487–496, 1997.)

Short TE spectra contain short T2 metabolite peaks that are not visible on long TE sequences. The most prominent of these is at 3.56 ppm and represents the methine protons on C1, C3, C4 and C6 in the myoinositol (mI) molecule. The mI peak is dominant in a normal newborn brain spectrum, but decreases during brain development (Fig. 6.7b). Myoinositol increases with

Alzheimer's disease[47,48] and diabetes mellitus,[49] and decreases in stroke, tumor, and hypoxic encephalopathy; it has been shown to be depleted from the brain spectra of patients with chronic hepatic encephalopathy.[11,12] Another smaller mI resonance, from methine protons on C2, is located at 4.06 ppm and may be seen with good water suppression.

Other signals seen on short TE include glucose, GABA, and glutamine (Gln) and glutamate (Glu), commonly referred to as Glx peaks. Protons of the β-CH$_2$ and γ-CH$_2$ groups of Glx produce peaks in the 2.0–2.45-ppm range, and protons of the α-CH group produce signals in the 3.6–3.8-ppm range. These are very small peaks due to complex coupling processes and fast T2 decay. They are hardly seen in normal brain spectra but become more visible as a result of elevated glutamine in brain injury or some encephalopathies. The Glx peaks increase in cases of near drowning, hypoxia, and hepatic[11,12] and hypoxic[27] encephalopathies. Glucose (Glc) has peaks at 3.43 and 3.80 ppm. The glucose peaks from ^{1}H protons on C1–C6 are elevated in brain spectra of patients suffering from diabetes.[48] Gamma-aminobutyric acid (GABA) has peaks at 1.9 and 2.3 ppm from protons in the β- and α-CH$_2$ groups, and at 3.00 ppm from the γ-CH$_2$ group. This latter peak is usually hidden by the Cr + PCr peak that occurs at the same location but may be detected using spectral editing techniques that remove the Cr peak. The GABA concentration has been shown to increase with the administration of vigabatrin, an antiepileptic medication.[50–52] Suggestions have been made to use GABA for monitoring patients with epilepsy that are being treated with vigabatrin.

Mobile lipid signals resulting from pathology appear as sharp peaks at 0.9 and 1.3 ppm and interfere with lactate when it is also present. Lipid signals are more prominent with short TE sequences. They are seen in some tumors, stroke, and acute MS lesions, and appear to be associated with acute destruction of myelin.

It is also possible to observe peaks in brain spectra that are not part of normal or diseased brain metabolism. For example, an ethanol signal, which was attributed to presumed ingestion of alcohol immediately prior to the MRS exam, has been observed as a triplet at 1.2 ppm and is described in Ross et al.[5]

Several studies have observed regional variations in normal brain metabolites as well as variations with age,[19,21,22,53–56] especially in the developing brain (Fig. 6.7) (Table 6.4). Metabolite concentrations in the developing brain,[53] calculated from short TE (30 ms) spectra, show that the dominant peak in a human brain spectrum changes from being mI in neonates, to being Cho in older infants, and then to NAA in adults. These metabolite changes are most rapid during the first 6 months and begin to level off at about 30 months of age (Fig. 6.8). They consist of a decrease in mI and Cho levels,

TABLE 6.4. ¹H Brain Metabolite Ratios in Neonates, Infants, and Children[a]

Ratio	Neonates ($\leq$ 1 mo) ($n = 3$)	Infants (1–8 mo) ($n = 15$)	Children ($n \geq 18$ mo) ($n = 6$)
NAA/Cr	0.84 ± 0.28	1.2 ± 0.17	1.53 ± 0.13
NAA/Cho	0.67 ± 0.17	1.56 ± 0.67	3.43 ± 1.04
Cho/Cr	1.27 ± 0.40	0.84 ± 0.21	0.49 ± 0.16

[a]Metabolite ratios calculated from peak areas measured in the occipital gray matter with a STEAM sequence (TR = 3000 ms, TE = 20 ms). Adapted from Holshouser et al.[56]

and an increase in NAA and Cr. The NAA and Cr are at their highest concentrations in the adult brain. Other studies concluded that metabolite variations in adulthood are marked by decreased NAA, Cr and Cho levels in older adults compared to young adults[4,20,21,22] (Table 6.5).

Despite some differences among studies in absolute metabolite concentrations, data from an individual study reflect regional variations in normal adult brain. Michaelis et al.[14] report absolute metabolite concentrations in the parietal white matter, parietal gray matter, cerebellum, thalamus, and pons. The data (Table 6.6) show that the concentrations of NAA, Cr + PCr, and mI are higher in gray than in white matter. Other published measurements[15] of absolute metabolite concentrations in white and gray matter of adult brain have similar trends with slightly different values (Table 6.7). Additional studies based on CSI measurements[58,59] have confirmed the presence of metabolic heterogeneity in normal brain tissue. A recent investigation[59] in the cortex and neocortex brain areas of healthy elderly volunteers using a multislice CSI technique showed that a progressive decrease in gray matter NAA from frontal and parietal to temporal lobes, with a lower concentration in white matter and still lower concentration in the hippocampus.

No differences in metabolite concentrations have been reported between equivalent locations in the right and left hemispheres of normal brain. This observation makes it possible to examine focal disease using contralateral spectra of normal brain tissue as reference. Initial reports[22,60] have indicated no gender dependence of metabolite peaks or ratios for the same region in the brain. A recent study[61] shows differences in NAA/Cho and Cho/Cr between male and female groups. The reported NAA/Cho is 2.37 ± 0.4 for men and 2.78 ± 0.45 for women while Cho/Cr is 1.05 ± 0.23 for men and 0.92 ± 0.16 for women.

Some of the reported ¹H metabolic changes associated with brain disease are listed in Table 6.8.[4,6] Figures 6.9–6.22 show spectra that illustrate certain metabolite peak variations with disease. (See color section at end of this chapter for Figures 6.13, 6.14, and 6.19.)

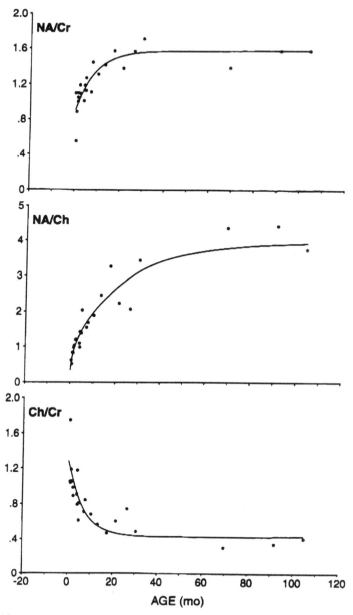

Figure 6.8. Graphs show age-dependent changes in NAA/Cr (NA/Cr), NAA/Cho (NA/Ch), and Cho/Cr ratios in occipital gray matter in 24 developmentally normal control patients. Solid lines represent fitted curves. [Reproduced by permission of RSNA Publications from B. A. Holshouser, et al., *Radiology, 202*, 487–496, 1997.]

TABLE 6.5. ¹H Metabolite Ratios in Normal Adult Brain[a]

	White Matter			Gray Matter		
Age (yrs)	NAA/Cr	Cho/Cr	mI/Cr	NAA/Cr	Cho/Cr	mI/Cr
16–25	1.54 ± 0.09	0.77 ± 0.05	0.59 ± 0.03	1.41 ± 0.08	0.56 ± 0.05	0.6 ± 0.04
26–37	1.49 ± 0.07	0.78 ± 0.06	0.60 ± 0.03	1.36 ± 0.08	0.61 ± 0.08	0.6 ± 0.04
40–78	1.41 ± 0.12	0.82 ± 0.02	0.63 ± 0.04	1.26 ± 0.09	0.60 ± 0.05	0.59 ± 0.06

[a]Adapted from Ross and Michaelis.[4]

TABLE 6.6. Absolute Concentrations of Normal Adult Brain ¹H Metabolites[a]

Metabolite	White Matter (parietal)	Gray Matter (parietal)	Cerebellum (central)	Thalamus	Pons
NAA + NAAG	11.2 ± 1.2	11.7 ± 2.2	9.6 ± 1.6	11.8 ± 0.7	12.5 ± 0.8
NAA	8.8 ± 1.0	11.7 ± 2.2			
PCr + Cr	6.1 ± 0.8	8.2 ± 1.4	9.1 ± 1.2	7.5 ± 0.6	6 ± 0.3
Cho	1.8 ± 0.3	1.4 ± 0.3	2.5 ± 0.3	1.8 ± 0.2	2.9 ± 0.7
mI	4.7 ± 1.0	6.2 ± 1.1	6.8 ± 1.3	4.9 ± 0.6	5.4 ± .06
Glu	8.1 ± 1.5	12.5 ± 3	9.5 ± 2.1		
Glc	1.0 ± 0.2	1.1 ± 0.4	1.2 ± 0.4		

[a]Concentrations are in mM, corrected for coil loading, T2 effects and partial volume effects. Adapted from Michaelis et al.[14]

TABLE 6.7. Absolute Concentrations of Normal Adult Brain ¹H Metabolites[a]

Metabolite	White Matter	Gray Matter
NAA	8.8 ± 0.18	9.24 ± 0.26
PCr + Cr	6.33 ± 0.14	8.04 ± 0.32
Cho	1.58 ± 0.02	1.43 ± 0.08
mI	6.68 ± 0.44	7.26 ± 0.52

[a]Concentrations are in mM, corrected for coil loading, T2 effects and partial volume effects. Adapted from Kreis et al.[15]

TABLE 6.8. Variations of Major ¹H Brain Metabolites in Selected Diseases[a]

Disease	Metabolic Changes					References
Stroke	↓NAA	↓Cr[b]	↓Ins	↑↑Lac		36–38
Glial tumors	↓NAA ↓Cr	↑Cho	↑lac	↑Lipids (not always)		31, 62, 63, 64
Meningiomas	↓NAA ↓↓Cr	↑Cho	↑alanine	↑aspartate[c]		31, 45, 46
Necrosis[d]	↓↓↓NAA	↓↓↓Cr	↓↓↓Cho			65, 66
HIV and related diseases	↓NAA	↓Cr?	↑Cho	↑lac		67–71
Multiple sclerosis	↓NAA ↑Cho	↑lac? (active plaques)	↑Lipids			72–76
Alzheimer's	↓NAA	↑Cho	↑Ins			20, 32–34, 48
Epilepsy	↓NAA	↑Cho				16, 35, 77
Hepatic encephalopathy	↓Cho	↓Ins	↑Glx			11, 12
Hypoxia	↓NAA ↓Cr ↑Cho (chronic hypoxia) ↑Glx ↑lac ↑Ins (recovered hypoxia)					4, 27
Anoxia	↓NAA	↑lac				4
Brain abcess	Absence of NAA, Cho and Cr. ↑Ace[e] ↑Lac ↑Suc[e] ↑AA[e] ↑lipids					78
Diabetes Mellitus	↓NAA	↑Glc	↑Cho	↑Ins		49

[a]Adapted from Ross and Michaelis[4] and Vion-Dury et al.[6] and updated.
[b]Cr refers to total creatine (Cr + PCr)
[c]Aspartate has a chemical shift of 2.8 ppm.
[d]Necrotic tissue is characterized by the absence of metabolites. The metabolite peaks sometimes seen on spectra of necrotic tissue are due to contamination from normal brain tissue included in the measured volume.
[e]Ace is acetate (1.9 ppm). Suc is succinate (2.4 ppm). AA represents amino acids that include valine (0.9 ppm), alanine (1.5 ppm), and leucine (3.6 ppm).

6.3.4. MRS Following Contrast Agent Administration

The effect of a contrast agent on the outcome and interpretation of MRS spectra is a critical issue in view of the fact that, in some brain pathologies, contrast enhanced MR images are of primary importance and take precedence over an MRS exam. The contrast agent may shorten the T1 relaxation time of the metabolites, increasing their signal for short TR scans. The local

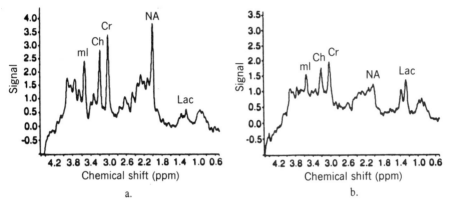

Figure 6.9. Serial spectra obtained in occipital gray matter in a 2-year-old girl after cardiac arrest due to accidental hanging. Proton MR spectra (3000/20 ms, STEAM) obtained at (*a*) 36 h and (*b*) 4 days after injury show the progressive decrease in NAA (NA) and increase in lactate (lac) with increasing time after injury. The patient was comatose with reactive pupils; her condition evolved to and remains a permanent vegetative state. The presence of lactate was confirmed at 36 h with a long-echo-time (270 ms) acquisition (not shown). [Reproduced by permission of RSNA Publications from B. A. Holshouser et al., *Radiology, 202,* 487–496 (1997).]

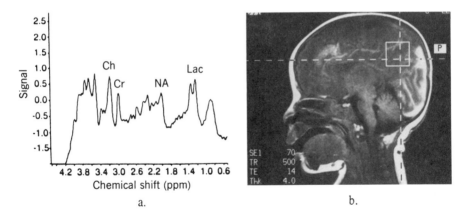

Figure 6.10. Proton MR spectrum in a 5-week-old infant who experienced seizures, lethargy, and apnea due to nonaccidental trauma illustrates the importance of the presence of lactate as a prognostic indicator in traumatic brain injury. (*a*) Proton MR spectrum (3000/20 ms, STEAM) obtained on day 4 after injury reveals a prominent lactate (lac) peak with NAA (NA)/Cr ratio reduced more than 2 SD below normal. (*b*) Sagittal spin echo image also obtained on day 4 shows an extensive occipital pole subarachnoid hemorrhage. Other findings included diffuse cerebral edema, subfalcial herniation, and right frontal subdural hematoma. The square represents the localized region of interest used at ¹H MR spectroscopy. The patient's condition evolved to a severely disabled state. [Reproduced by permission of RSNA Publications from B. A. Holshouser et al., *Radiology, 202,* 487–496 (1997).]

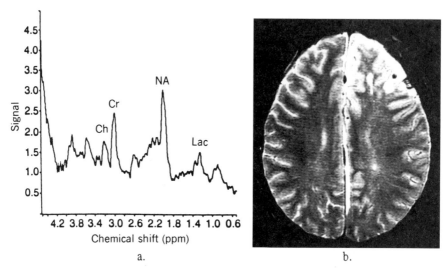

a.   b.

Figure 6.11. Proton MR spectroscopy in a 15-year-old-boy with bacterial meningitis illustrates that the presence of lactate does not appear to correlate with a poor outcome in children CNS infections. (*a*) Proton MR spectrum (3000/20 ms, STEAM) obtained on day 6 after hospital admission shows a prominent lactate (lac) peak and normal NAA (NA)/Cr, NAA (NA)/Ch and Ch/Cr ratios. (*b*) Axial spin echo MR image also obtained on day 6 shows residual empyema with left frontal edema after surgery performed on day 1. By 12 months, the patient recovered to a good outcome. [Reproduced by permission of RSNA Publications from Barbara A. Holshouser et al., *Radiology, 202,* 487–496 (1997).]

tissue susceptibility may also be shortened by the agent, reducing T2* and causing line broadening. Recent investigations[79–82] have demonstrated that good quality spectra may be obtained on postcontrast MRS measurements. All studies have used the same gadolinium contrast agent and report that the Cho peak is broadened most on the postcontrast spectra. Some results show no quantitative differences between pre- and postcontrast spectra[80] collected with long and short TEs and a TR of 3000 ms; others have reported significant changes in metabolite ratios[81] or peak areas.[82] One study[82], using 2D CSI (TR = 1500 ms, TE = 135 ms) reports a 15% decrease in the Cho peak area. The loss in Cho signal is attributed to shortening of the metabolite T2* on postcontrast MRS. Another study[79] based on SVS with PRESS (TR = 1500 ms and TE = 135 ms), reports an increase in the peak areas of Cho (3%), Cr (20%), and NAA (20%). This increase is attributed to reduced T1 relaxation times for the metabolites on the postcontrast scan. These T1 effects may be removed with the use of a long TR that allows complete recovery of the magnetization from all metabolites. The discrepancies among the various studies may be due to differences in spectroscopic techniques and data analysis procedures, or to differences in brain tissue abnor-

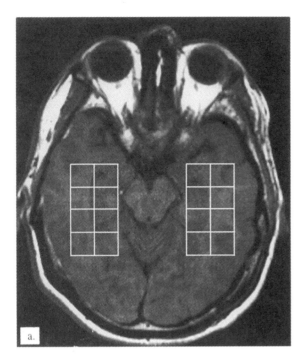

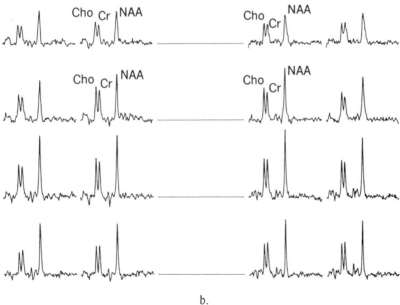

b.

Figure 6.12. (*a*) Positioning of the VOIs for a temporal lobe epilepsy MR examination. (These VOIs are simulated on the head image of a volunteer.) (*b*) Spectral maps from VOIs similar to the ones depicted in a). Despite regional variations, epileptogenic index NAA/Cho, is always visually greater than one. Figures 6.13 and 6.14 show data from patients with temporal lobe epilepsy. [Spectra: courtesy of T. C. Ng, Cleveland Clinic Foundation.]

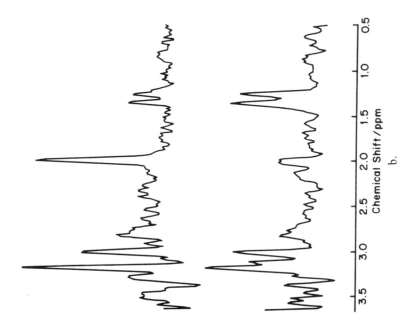

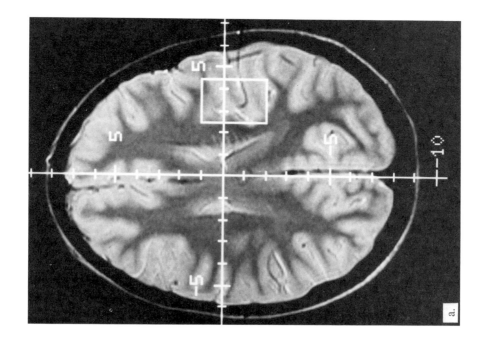

Chemical Shift / ppm

b.

a.

malities and related variations in relaxation times. In view of the many factors affecting quantitation, more investigations are needed for accurate evaluation and interpretation of postcontrast spectra. Nonetheless, these may be used when only qualitative analysis is desired or when quantitation is not critical to the interpretation of spectral data. Postcontrast MRS has been recommended for better visualization of conspicuous lesions and more accurate definition of the VOI.

6.3.5. ¹H MRS in Other Organs

Pilot studies of ¹H MRS outside the brain have been reported in muscle, bone marrow, prostate, breast, and liver. Good quality high-resolution ¹H MRS spectra from the body are difficult to produce because of inhomogeneous line broadening, large lipid signals, and physiological motion. The dominant peaks are usually lipids and water with typical line widths on the order of 12 Hz. Initial analysis[83] of the fat signal in a human leg muscle showed differences between individuals with neuromuscular diseases compared to healthy volunteers. The major peaks in skeletal muscle[84] are triglicerides, betaines (Cho-containing compounds, carnitines) and phosphocreatine, with a small contribution from creatine. Comparison of signals from fat in skeletal muscle tissue and in fat tissue revealed that there are two fat compartments in muscle.[85] It has been suggested that the characteristics of fat peaks and the relative water/fat concentration be tested for the assessment of muscle disease. More recently,[86] ¹H chemical shift imaging has been used to quantify and image the total creatine in human calf muscle, which will enable a better understanding of the creatine metabolism. The ability to quantify creatine levels in muscle will assist in the evaluation of the response to creatine therapy.

¹H MRS has also been used to examine lumbar vertebral bodies[87] and bone marrow fat in patients with leukemia[88] for monitoring the effect of chemotherapeutic treatment. The only peaks seen in spectra of bone marrow are water and lipids. It was found that iliac bone marrow in patients with

---

Figure 6.15. A 17-year-old girl with diabetic ketoacedosis. (*a*) Initial MR image shows no apparent abnormality. Spectra are acquired in region indicated by box. (*b*) The MR spectra were acquired with simulated-echo acquisition mode (STEAM) sequence (TR/middle interval/TE, 1500/30/270 ms). Upper trace: spectrum obtained during initial examination. Lactate (1.3 ppm) is elevated although findings on MR image were normal. Lower trace: spectrum from same region 1 week later. Lactate level was much higher and level of *N*-acetyl asparate (2.02 ppm) was drastically reduced. MR imaging finding of area remain normal (not shown). [Reproduced by permission of ARRS publications from Z. Wang et al., *AJR, 167,* 191–199 (1996).]

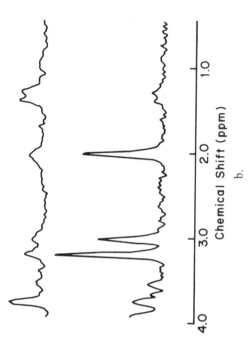

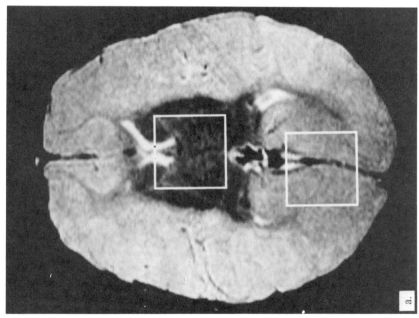

Chemical Shift (ppm)

4.0   3.0   2.0   1.0

leukemia has a lower fat content and a longer water T1 compared to healthy volunteers. Another study[89] found that biopsy results correlated well with the assessment of bone marrow cellularity measured from the relative fat/water concentration in ^{1}H MRS spectra.

MRS has been combined with MRI examinations of the breast in an attempt to improve the diagnostic utility of breast MRI. The dominant peak in normal ^{1}H MRS spectrum of breast tissue is from fat, with a water/fat ratio of about 0.3. Initial studies of the breast revealed a decrease in the water/lipid ratio in breast carcinoma compared to healthy breast tissue. A recent localized ^{1}H MRS study,[90] using specially designed breast coils, investigated the Cho peak and reported differences in Cho levels between malignant and benign breast tumors. Further research is needed in order to determine the usefulness of the Cho peak in the characterization of small breast lesions.

Several ^{1}H MRS studies[91–95] have reported variations in citrate (and other metabolites) levels that may be used to differentiate between benign disease and malignant prostatic tumors. It is known from biochemical analysis that the concentration of citrate is elevated in benign prostatic hyperplasia compared to normal prostate tissue and is decreased in prostatic cancer. Good quality ^{1}H MRS spectra and well-resolved citrate peaks can be obtained using either an endorectal coil or phased array surface coils.[91] Improved measurement techniques consist of combining water and fat suppression with high resolution 3D CSI and quantitation that takes into account the receiving profile of a surface coil.[92] These are being employed in the assessment of prostate tissue including evaluation of response to therapy and differentiation between residual cancer and normal tissue. Very promising results have been reported from combining MRS measurement of citrate and choline to an MRI examination of the prostate gland; while citrate is decreased with cancer, Cho is elevated and the citrate/Cho + Cr ratio seems to provide a good indicator of cancer versus normal tissue.

Figure 6.16. A 2.6-month-old girl with cerebral infarct resulting from head trauma. (*a*) T2-weighted MR image shows extent of infarct after injury. Most of the brain is swollen except for ganglionic region at center. (*b*) Results of MR spectroscopy. Upper trace: spectrum from infarct in occipital lobe. Lower trace: spectrum from uninvolved region in central part of the brain. Both MR spectra were obtained with simulated-echo acquisition mode (STEAM) sequence (TR/middle interval/TE, 1500/30/270 ms). Both spectra are scaled according to size of region of interest and plotted using same scale. In trauma area, lactate level (1.3 ppm) was increased, choline (3.22 ppm), creatine (3.02 ppm), and *N*-acetylaspartate (2.02 ppm) levels were decreased. Lactate level was also somewhat increased in control region. Part of this increase may be explained by hyperventilation, as patient was on life support. [Reproduced by permission of ARRS publications from Z. Wang et al., *AJR, 167*, 191–199 (1996).]

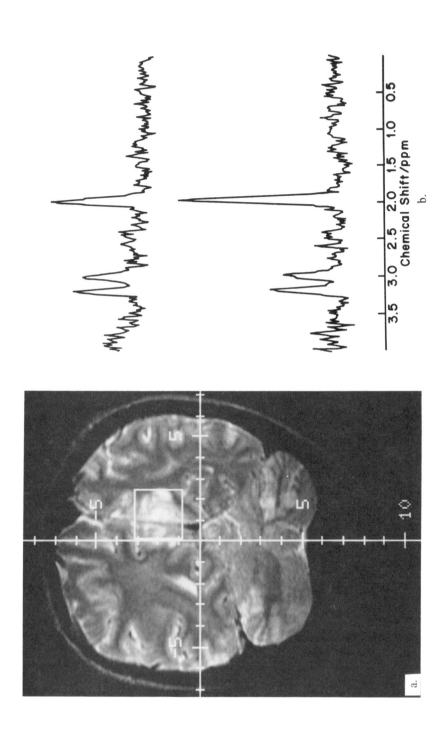

Chemical Shift/ppm

b.

a.

Because of the technical difficulties related to ^{1}H MRS in the body, in vivo ^{1}H MRS of the liver has not been employed as much as ^{31}P MRS. An investigation of fatty infiltration[96] with ^{1}H MRS found a good correlation between the assessment of hepatic fat by MR and that by computed tomography density measurement.

6.4. PHOSPHORUS SPECTROSCOPY

Early attempts at in vivo MRS focused on the development of phosphorus spectroscopy because of its potential in the investigation of cellular energy metabolism. Phosphorus has a larger chemical shift range ($\sim$30 ppm) than hydrogen ($\sim$4 ppm) and a much lower sensitivity. Its most common applications are in the brain, muscle, liver, and heart. Some studies have been reported on kidneys, prostate, and other in vivo tissues. Solvent suppression schemes are not required in ^{31}P MRS, but lower spatial resolution and/or increased exam times are required to achieve adequate S/N.

The major metabolite peaks that can be detected with phosphorus spectroscopy are (Fig. 6.23 and 6.24): the three peaks of adenosine triphosphate seen at -16.3 (β-ATP, triplet), -7.8 ppm (α-ATP, doublet), and -2.7 ppm (γ-ATP, doublet); phosphocreatine[f] (PCr) at 0 ppm; phosphodiester compounds (PDE) at 2.6 ppm, inorganic phosphate (Pi) at 4.9 ppm, and phosphomonoester compunds (PME) at 6.5 ppm. Cellular energy metabolism is represented by ATP, PCr, and Pi, with PCr serving as a high-energy phosphate storage compound in certain tissues such as muscle and brain. The phosphodiesters and phosphomonoester compounds are from membrane phospholipids. An increase in PME has been associated with rapid tissue growth or rapid membrane synthesis. PME has also been found to be elevated in infant brain and in tumors. The chemical shift of the Pi peak is pH

[f]For practical reasons, the PCr peak in clinical ^{31}P MRS spectra is used as a reference and set to 0 ppm (Fig. 6.24). The chemical shift of PCr with respect to phosphoric acid, a commonly used standard in ^{31}P MRS, is -2.52 ppm (Fig 6.23).

Figure 6.17. Acute stroke in a 15-year-old girl with sickle cell disease. At initial examination, patient had clinical symptoms of acute stroke. Patient recovered quickly and follow-up study was done 4 weeks later. (*a*) Initial T2-weighted image shows small area of ischemia in occipital lobe. (*b*) Upper trace: spectrum of acute stroke at first examination. *N*-acetyl aspartate (2.02 ppm) level is decreased. Lower trace: spectrum from same region of interest 4 weeks later. Both spectra were acquired with simulated-echo acquisition mode (STEAM) sequence and identical pulse parameters (TR/middle interval/TE, 1600/30/270 ms) and plotted on same scale. Second MR spectrum shows level of *N*-acetylaspartate has returned to higher value. [Reproduced by permission of ARRS publications from Z. Wang et al., *AJR*, *167*, 191–199 (1996).]

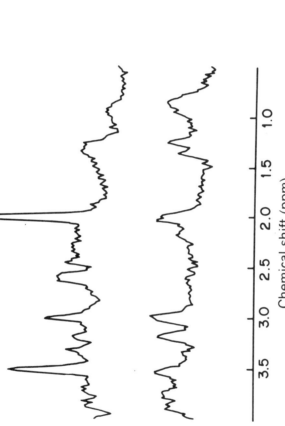

Figure 6.18. The MR spectra of leukodystrophy. Upper spectrum: Canavan's disease in 13-month-old girl; lower spectrum: metachromatic leukodystrophy in a 15-month-old girl. Both spectra were acquired with simulated-echo acquisition mode (STEAM) sequence (TR/middle interval/TE, 1600/30/270 ms) from periventricular white matter. [Reproduced by permission of ARRS publications from Z. Wang et al., *AJR, 167,* 191–199 (1996).]

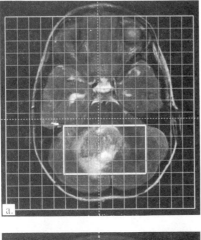

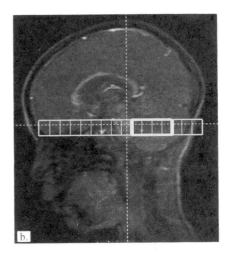

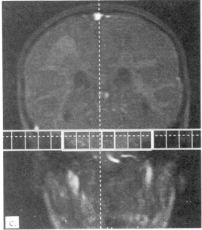

Figure 6.20. Transverse (*a*), sagittal (*b*), and coronal (*c*) images showing the positioning of a chemical shift imaging VOI on a cerebellar primitive neuroectodermal tumor. The corresponding spectral data are presented in Figures 6.21 and 6.22. [data: courtesy of Z. Wang, The Children's Hospital of Philadelphia, Philadelphia, PA.]

dependent.[g] It is commonly used for in vivo measurement of intracellular pH.[97] Normal tissue pH corresponds to a shift of 4.9 ppm.

6.4.1. Patient Exam:

Phosphorus spectroscopy requires double resonant coils that can operate at the hydrogen and phosphorus frequencies. Those may be either volume coils for head studies, or surface coils for muscle, liver, and heart studies. The

---

[g]Refer to Section 2.6.2 for more details.

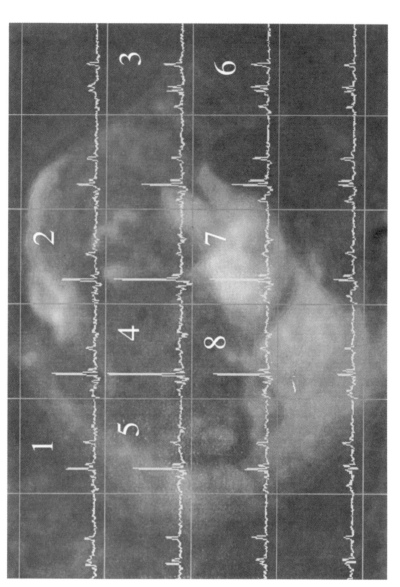

a.

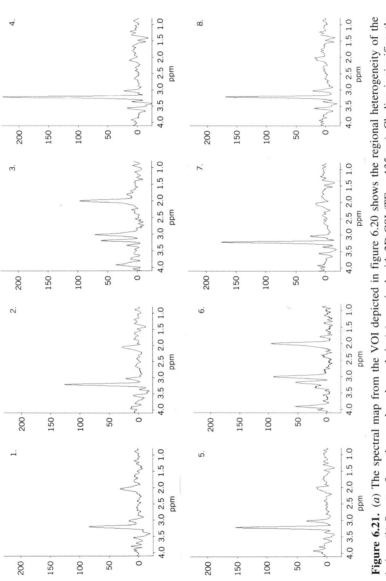

Figure 6.21. (*a*) The spectral map from the VOI depicted in figure 6.20 shows the regional heterogeneity of the lesion. (*b*) Spectra from the numbered voxels in (*a*) acquired with 2D CSI (TE = 135 ms). Choline is significantly elevated in 1, 2, 4, 5, 7, and 8. [Raw data: courtesy of Zhiyue Wang, The Children's Hospital of Philadelphia, Philadelphia, PA.]

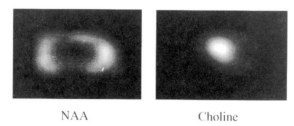

NAA   Choline

Fig. 6.22. Metabolite images indicate depletion of NAA and increase in choline at the center of the lesion described in Figures 6.20 and 6.21. [Raw spectral data: courtesy of Z. Wang, The Children's Hospital of Philadelphia, Philadelphia, PA.]

hydrogen frequency is used for shimming and for acquiring images. The phosphorus frequency is for collection of the MRS data. The general approach to a ^{31}P MRS patient examination is the same for any part of the anatomy. It consists of

- Patient positioning.
- Global shimming.
- Acquisition of MR images for localization.
- Selection of the ^{31}P MRS measurement sequence and parameters.
- MRS data collection.
- MRS data processing.

The patient is positioned such that the anatomy to be examined is centered both in the coil and in the magnet. For heart or liver studies, the surface coil is placed in front of the body. A prone position has the advantage of placing the organs to be examined closer to the coil surface. With the scanner operating at the hydrogen frequency, shimming is performed on the water signal and MR scout images are obtained. The scanner is then switched to the phosphorus frequency. A ^{31}P MRS sequence, typically CSI, is selected and the VOI is defined on the MR images. The MRS data are then collected and processed. For ^{31}P MRS measurements from the heart, triggering is incorporated in the measurement sequence to reduce motion artifact. For liver studies, saturation regions may be placed over moving anatomical parts such as the abdominal wall.

6.4.2. ^{31}P MRS in the Brain

A normal phosphorus brain spectrum contains seven peaks: the three ATP[h] peaks, PCr, PDE, Pi, and PME (Fig. 6.23). Metabolic changes in the develop-

[h]Because adenosine triphosphate (ATP) resonates at the same frequencies as other nucleoside triphosphates (Fig. 2.16), the ATP peaks are sometimes labeled NTP as in Figure 6.23.

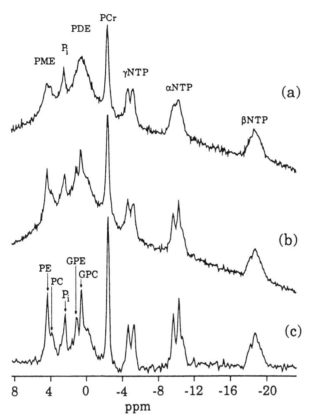

Figure 6.23. Nonlocalized ^{31}P spectra from the entire head displayed with the same vertical scale: (*a*) coupled and (*b*) proton decoupled, processed with no line broadening, and (*c*) the proton-decoupled data of (*b*) processed using 200-Hz deconvolution (0.8 subtraction factor) followed by 6 Hz Gaussian and −5 Hz exponential filters. [Reproduced by permission of Wiley publications from J. Murphy-Boesch et al., *NMR Biomed., 6,* 173–180, 1993.]

ing brain have been investigated with ^{31}P MRS.[19,98–101] At birth and for the first 3 years, PME is elevated while PDE is low; PME decreases and PDE increases from birth until they level off after age three. No changes were observed in PCr or ATP during the same period of time. Some of the reported phosphorus metabolic changes associated with brain disease are listed in Table 6.9.[6]

6.4.3. ^{31}P MRS in Muscle

One of the first applications of phosphorus spectroscopy was the investigation of muscle metabolism and changes related to exercise, fatigue, and disease. The normal phosphorus spectrum of muscle (Fig. 6.24) at rest contains the three ATP peaks, PCr, PME, Pi, and PDE, with PCr being the

TABLE 6.9. Variations of Major ^{31}P Brain Metabolites in Selected Diseases

Disease	^{31}P Metabolic Changes	Reference
Stroke	↓ PCr ↓ ATP ↑ Pi ↓ pH followed by ↑ pH	102–105
Malignant glial tumors	↓ PCr ↓ PDE ↑ Pi ↑ PME ↑ pH	31, 105, 107
Meningiomas	↓ PCr ↓ PDE, ↓ Pi ↑ pH	31, 106
HIV and related diseases	↓ PCr ↓ ATP ↑ Pi ↑ PME	108, 109
Multiple sclerosis	↑ PCr	110, 111
Alzheimer's disease	↑ PCr ↑ PDE ↓ Pi ↑ PME	112–118
Epilepsy	↓ PCr ↑ Pi ↓ PME ↑ pH	119–123
Migraine	↓ PCr; ↑ ADP ↑ Pi	124–127

^{a}Adapted from Vion-Dury et al.[6]

dominant peak. Inorganic phosphate and adenosine diphosphate (ADP) are the products of the ATP breaking down to produce the energy needed for muscular contraction.

$$ATP \rightarrow ADP + Pi + energy$$

While the·ATP concentration is maintained during aerobic respiration through glycolysis and oxidative phosphorylation, it is maintained during short-term anaerobic respiration through the breakdown of phosphocreatine via creatine kinase:

$$PCr + ADP \leftrightarrows ATP + Cr$$

In the evaluation of muscle spectra typically obtained with surface coils, β-ATP is used as an internal reference based on the assumption that the ATP concentration is constant. The α- and β-ADP peaks have the same chemical shifts as the α- and γ-ATP peaks, respectively. This overlap makes them unsuitable for such a reference. During steady-state aerobic exercise, PCr decreases while Pi increases (Fig. 6.24). The correlation between the Pi/PCr ratio and the rate of work is known as the transfer function[128] and has been used in the assessment of various muscle conditions. After exercise, there is a sharp increase in PCr followed by a slower recovery. The rate of PCr recovery has been found to depend linearly on the intracellular pH.[129] Examples of spectral abnormalities in muscle are a high Pi/PCr ratio at rest with a

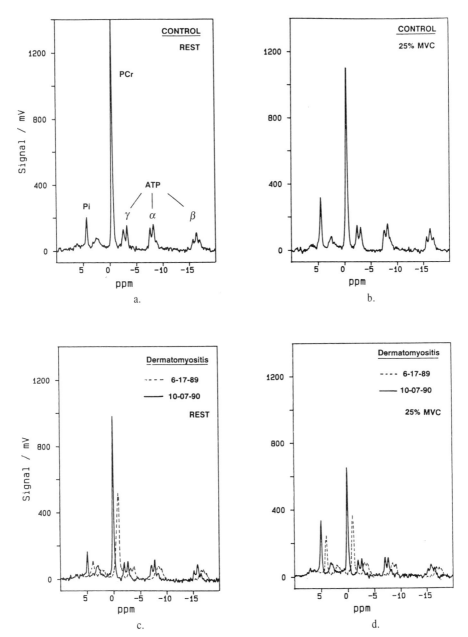

Figure 6.24. The ^{31}P spectra of quadriceps muscles from: (*a*) A normal control subject at rest. (*b*) A normal control subject during exercise. (*c*) A patient with dermatomyositis at rest. (*d*) A patient with dermatomyositis during exercise. The muscles of extremely weak patients had much lower levels of PCr and ATP during rest and exercise than did those of the control subjects. [Data courtesy of Jane Park, Vanderbilt University Medical Center, Nashville, TN.]

slightly alkaline pH in Duchenne muscular dystrophy; and a slow recovery of PCr after exercise with an abnormal transfer function[130] in mytochondrial myopathies.[131,132] Serial ^{31}P MRS examinations and measurements of the Pi/PCr ratio are being used to monitor the effects of treatment on patients with dermatomyositis.[133]

6.4.4. ^{31}P MRS in the Heart

Cardiac MRS is the most challenging in vivo spectroscopy application in terms of data acquisition and data analysis. The purpose of MRS measurements is to assess the function of the myocardium through variations in high-energy phosphates. Difficulties in the spectroscopy measurements stem from the heterogeneity and motion of the heart, techniques that can only measure the anterior part of the myocardium, and contamination of the myocardial spectrum with signals from blood and chest muscle. Most early cardiac MRS data were obtained using surface coil localization techniques[i] that collected spectra from an undefined volume, or from defined slices parallel to the surface of the coil. Newer techniques such as 2D and 3D CSI allow localization to small voxels that may be 8–27 cm^3. The spectrum of a normal heart contains the three ATP peaks, a narrow PCr peak, a smaller PDE signal, and, in case of contamination, two peaks at 5.4 and 6.3 ppm from blood 2,3-diphosphoglycerate (DPG). The Pi peak is not always resolvable in heart spectra due to the overlap with the DPG peaks, which interferes with intracellular pH measurements. This problem is alleviated by measurement sequences[j] that eliminate signals from the blood.[134,135] The relevant ratios for assessing the energy metabolism are PCr/ATP and PCr/Pi. Numerous measurements of PCr/ATP in normal heart have been reported in the literature. These measurements show a wide range of variability due to the variety of techniques, blood contamination, and saturation effects. With the appropriate corrections taken into account, it was deduced that for normal heart, PCr/ATP has a value of 1.83 ± 0.12.[136,137] Investigations into a correlation between human heart disease and variations in the concentration of high energy metabolites suggest that PCr/ATP is reduced only in very severe heart failure. In an experiment using hand grip exercise,[136] the myocardium PCr/ATP in normal volunteers was not changed during exercise. In patients with irreversible ^{201}Tl perfusion defect, the ratio decreased with exercise and then remained constant; in patients with reversible ^{201}Tl defect, the ratio was measured to be normal at rest and decreased during exercise. Although other trends of variations in high-energy phosphate with heart disease have been documented, there are discrepancies in the reported results

---

[i]Surface coil localization is described in Chapter 4 and Appendix A.
[j]Currently, those sequences are not available commercially.

and more data are needed in order to establish the clinical utility of ^{31}P MRS in the heart. Another important area for investigation is the noninvasive monitoring of cardiac transplant rejection by ^{31}P MRS. Studies in rats have shown that PCr/ATP, PCr/Pi, and ATP/Pi ratios are significantly reduced in cardiac rejection.[138] Studies on humans[139] have reported significant differences in the PCr/ATP ratio of patients with mild and moderate rejection. However, another study[140] concluded that ^{31}P MRS was not able to precisely predict histologic rejection in many patients with heart transplants. Further studies are needed in order to clarify the role of ^{31}P MRS in monitoring heart transplant patients.

6.4.5. ^{31}P MRS in Other Organs

A distinctive feature of phosphorus spectra from organs other than brain and muscle is that they do not contain a PCr peak due to the absence of creatine kinase activity. The PCr peak that is sometimes seen, for example, in liver spectra, indicates contamination from nearby muscle. Liver ^{31}P MRS is typically performed with surface coils and a variety of techniques including DRESS,[k] ISIS, 1D, and 2D CSI.[l] Because measurement techniques and parameters affect signal intensities differently, there are discrepancies in the published metabolite ratios in normal liver. Nonetheless, relative results from normal and abnormal liver at a given center are clinically useful. One of the applications of phosphorus MRS in the liver is the assessment of its ability to metabolize fructose.[141] It was suggested that constant fructose infusion combined with phosphorus MRS may serve for the evaluation of liver function. Among the liver diseases[142–145] that were investigated with ^{31}P MRS are tumor, hepatitis, and cirrhosis. Carcinoid metastases have shown elevated levels of PME and a decrease in PDE/ATP ratio. An investigation[144] of primary and metastatic liver tumors revealed an elevated PME/ATP ratio with low ATP, while some tumors had low Pi levels. Spectral variations with hepatitis consist of increased PME and reduced PDE. In liver cirrhosis[145] there is an increase in PME/ATP, a decrease in PDE/ATP, and a low pH that may be the result of lactic acidosis. Phosphorus MRS may play a role in the assessment of diffuse liver disease and large tumors. It is currently investigational and not routinely used in clinical practice.

Phosphorus spectroscopy has been used to investigate prostate and breast tumors and to evaluate their response to therapy.[146–148] Pilot studies of prostate adinocarcinomas have shown an increase in the PME/β-NTP ratio and a reduction in the PCr/PME ratio. The ^{31}P spectrum of human normal breast tissue consists of low levels of PME, Pi, ATP, and PDE with some PCr

[k]See Appendix A for a description of DRESS.
[l]Refer to Chapter 4 for a description of techniques used.

observed in young volunteers. One study[146] on large breast tumors demonstrated high levels of PME and PDE. Another ^{31}P MRS study of breast cancer has shown[146] that the relative concentration of PME to γ-ATP increased in proliferating aneuploid tumors. The ratio decreased in response to therapy. The PME levels were also shown to decrease in response to tamoxiphen and cytotoxic chemotherapy. However, the small S/N ratio limits the usefulness of the technique to monitoring therapy in large tumors and does not help with early detection of small tumors.

Clearly, ^{31}P MRS can provide a wealth of information for the understanding and monitoring of energy metabolisms. Combined ^{1}H and ^{31}P MRS examinations provide complimentary metabolic information that increases the clinical utility of MR spectroscopy. Faster progress in MRS will result from improvements in coil technology such as phased array coils, better localization techniques, the ability to resolve and enhance phosphorus peaks using ^{1}H decoupling and NOE, and the design of reliable means for quantification.

6.5. CLINICAL APPLICATIONS OF MRS WITH OTHER NUCLEI

Magnetic resonance spectroscopy with nuclei other than hydrogen is purely investigational and mostly used at MRS research centers on animal models with some pilot studies on humans. Those nuclei include carbon, fluorine, lithium, şodium, potassium, and nitrogen.

The natural abundance of ^{13}C is 1.1%, and its MR sensitivity is 1.6% that of hydrogen. It can only be detected in storage compounds that have high intracellular carbon concentration, such as liver and muscle glycogen. The advantages of ^{13}C MRS are: spectra contain a large chemical shift range of about 200 ppm, a large number of metabolite peaks that are well separated, and the ability to enrich substrates that are metabolized in living cells for observing chemical pathways and measuring metabolic rates in vivo. The spin–spin J coupling between neighboring carbon and hydrogen nuclei in carbon-containing compounds causes the lines in a ^{13}C spectrum to split, thus complicating the spectrum and reducing the S/N ratio. Proton decoupling is commonly used with ^{13}C MRS to simplify spectra by collapsing multiple peaks into single ones and to increase the S/N ratio. The ^{13}C MRS studies of glycogen reveal well-resolved peaks implying that portions of this large molecule are highly mobile. Measurements of muscle glycogen have been shown to differentiate between patients having glycogen storage disease and normal volunteers.[149] Infusion of ^{13}C labeled glucose revealed that muscle glycogen synthesis provides the metabolic pathway for glucose disposal in normal and diabetic subjects.[150] Labeling is also employed in MRS studies of animal brain for elucidation of cerebral metabolic pathways such as the ones associated with lactate, glutamine, glutamate, and GABA synthesis. The drawbacks of ^{13}C MRS is the high cost of ^{13}C enriched compounds and the high rf power deposition associated with proton decoupling.

Fluorine has a natural abundance of 100% and a spin ½. Its MR sensitivity is 83% compared to hydrogen. These characteristics along with the fact that the endogenous concentration of fluorine in the human body is low make ^{19}F an excellent candidate for in vivo monitoring of fluorine containing drugs. The most promising clinical application of ^{19}F MR spectroscopy consists of the identification of tumor patients who respond to chemotherapy with 5-fluorouracil (5-FU). This monitoring helps in the optimization of individual treatments and allows better patient management.[151–153] Tumors that have been studied in this manner were located in the liver, lung, breast, and mediastinum. Fluorine spectra are easy to interpret because there are no complications from unwanted background signals and the peaks are well separated over a range of about 200 ppm.

Sodium has a natural abundance of 100% and an MR sensitivity of 13%; however, it has a spin of ³⁄₂, and the associated quadrupolar effects[m] result in partial detection of ^{23}Na. Interest in sodium stems from the major role it plays in the regulation of cellular function and its potential applications in assessing cell viability and restoration of normal cell function following damage. In normal cell metabolism, the concentration of intracellular ^{23}Na is 12 mM, whereas extracellular ^{23}Na has a concentration of 140 mM. An elevated intracellular sodium level is associated with cell damage such as ischemic or hypoxic insults. Therefore it is important to be able to monitor intracellular and extracellular sodium separately. In ^{23}Na MRS, the two sodium compartments produce indistinguishable MR signals that contribute to the same spectral peak. One technique for differentiating the two MR signals is to use shift reagents that do not penetrate cellular membranes, thus causing the resonant frequency of extracellular sodium to be shifted away from the resonant frequency of intracellular sodium. However, the clinical application of this technique is hampered by the toxicity of these reagents and by the difficulty in ensuring uniform dispersal into the extracellular space.

Potassium (^{39}K) is another nucleus that is involved in cellular function and that may be investigated in conjunction with sodium in studies of the Na–K activated ATPase. The natural abundance of ^{39}K is 93% and its MR sensitivity relative to hydrogen is 4.7%. Like sodium, it has a spin of ³⁄₂, a

---

[m]The nuclear quadrupolar interaction is described in Chapter 2.

quadrupolar moment, and short relaxation times; potassium is mainly intra-cellular.

Other nuclei that are being investigated for potential clinical applications include ^{17}O for the evaluation of ischemia and ^{7}Li for monitoring brain lithium in manic depressive patients that are undergoing therapy.

6.6. PAST, PRESENT, AND FUTURE

While imaging examinations have become progressively shorter as the tech-nology has evolved, there are two main obstacles that came in the way of a faster progress of clinical MRS. The first involved technical issues that made MRS too lengthy for a patient examination. The second is the lack of organized clinical trials on various pathologies and lack of established clini-cal guidelines to offer to the medical community. The second issue is a consequence of the first since few clinical MR practices can afford lengthy patient exams. An ^{1}H MRS of the brain could easily take 1 hour and the results were variable because of the sensitivity of the technique to prepara-tory adjustments. These adjustments consumed a significant fraction of the exam time and required scientific and technical knowledge. Current ad-vances in hardware and automated software have eliminated or reduced the tedious adjustment steps. Currently, ^{1}H SVS spectra in the brain can be obtained in 10 min or less, which is not much longer than imaging scans. MRS is at a stage now where the technological hurdle of running an exam and acquiring good data has been overcome; and with newer fast MRS sequences, and improved software the exam time will be reduced even further. The ease of performing a spectroscopy measurement is comparable to performing an imaging scan.

The next issue that needs to be settled consists of standardization of measurement and postprocessing techniques. The vast amount of MRS pub-lications, in the past, have covered a variety of techniques with a wide range of measurement parameters. This variety was expected in the experimental stage of MRS. It will continue to exist at investigational centers that will bring innovations to the field. However, for clinical viability, and in order to be able to compare patient studies from various clinical centers, there must be some standardization in measurement and data processing techniques. This standardization is starting to occur in the sense that there is a concerted effort to investigate ^{1}H MRS clinical applications in the brain with the same techniques and parameters. Data from normal tissue and from various patho-logies are being accumulated and compared. These gathered experiences will produce recommendations for clinical use, and, consequently, increase the utilization and usage of MRS.

REFERENCES

1. M. Paley: Proton spectroscopy of the human brain, in *Advanced MR Imaging Techniques,* William G. Bradley, Jr., and G. M. Bydder, eds. Martin Dunitz, pp. 309–332, 1997.

2. E. F. Jackson and C. A. Meyers: Introduction to hippocampal spectroscopy, *Neuroimaging Clin. N. Am.* **7(1)**, 143–154, 1997.

3. J. S. Ducan: Magnetic resonance spectroscopy, *Epilepsia* **37(7)**, 598–605, 1996.

4. B. Ross and T. Michaelis: Clinical applications of magnetic resonance spectroscopy, *Magn. Reson. Q.* **10(4)**, 191–247, 1994.

5. B. Ross, R. Kreis, and T. Ernst: Clinical tools for the 90s: magnetic resonance spectroscopy and metabolite imaging, *Eur. J. Radiol.* **14**, 128–140, 1992.

6. J. Vion-Dury, D. J. Meyerhoff, P. J. Cozzone, and M. W. Weiner: What might be the implication of the analysis of brain metabolites by in vivo magnetic resonance spectroscopy, *J. Neurol.* **241(6)**, 354–371, 1994.

7. M. Castillo, L. Kwock, and S. K. Mukherji: Clinical applications of proton MR spectroscopy, *AJNR* **17**, 1–15, 1996.

8. F. A. Howe, R. J. Maxwell, D. E. Saunders, M. M. Brown, and J. R. Griffith: Proton spectroscopy in vivo, *Magn. Reson. Q.* **9(1)**, 31–59, 1993.

9. J. P. Cousins: Clinical MR spectroscopy: Fundamentals, current applications, and future potential, *AJR* **164**, 1337–1347, 1995.

10. O. Henriksen: MR spectroscopy in clinical research, *Acta Radiologica* **35(2)**, 96–116, 1994.

11. R. Kreis, N. Farrow, and B. D. Ross: Diagnosis of hepatic encephalopthy by proton magnetic resonance spectroscopy, *Lancet*, 635–636, 1990.

12. R. Kreis, B. Ross, N. A. Farrow, and Z. Ackerman: Metabolic disorders of the brain in chronic hepatic encephalopathy detected with ^{1}H MR spectroscopy, *Radiology* **182**, 19–27, 1992.

13. T. Ernst and J. Hennig: Coupling effects in volume selective ^{1}H spectroscopy of major brain metabolites, *Magn. Res. Med.* **21**, 82–96, 1991.

14. T. Michaelis, K. D. Merboldt, H. Bruhn, W. Hanicke, and J. Frahm: Absolute concentrations of metabolites in the adult human brain in vivo: quantification of localized proton MR spectra, *Radiology* **187**, 219–227, 1993.

15. R. Kreis, T. Ernst, and B. D. Ross: Absolute quantitation of water and metabolites in the human brain. II. Metabolite concentrations, *J. Magn. Reson. B* **102**, 9–19, 1993.

16. T. C. Ng, Y. G. Comair, M. Xue, N. So, A. Majors, H. Kolem, H. Luders, and M. Modic: Temporal lobe epilepsy: presurgical localization with proton chemical shift imaging, *Radiology* **193**, 465–472, 1994.

17. J. Frahm, H. Bruhn, M. L. Gyngell, K. D. Merboldt, W. Hänicke, and R. Sauter: Localized high-resolution proton NMR spectroscopy using stimulated echoes: initial applications to human brain in vivo, *Magn. Reson. Med.* **9**, 79–93, 1989.

18. T. Michaelis, K. D. Merboldt, W. Hänicke, M. L. Gyngell, H. Bruhn, and J. Frahm: On the identification of cerebral metabolites in localized ¹H NMR spectra of human brain in vivo, *NMR Biomed.* **4**, 90–98, 1991.

19. M. S. Van der Knapp, J. van der Grond, P. C. Van Rijen, J. A. J. Faber, J. Valk, K. Willemse: Age-dependent changes in localized proton and phosphorus MR spectroscopy of the brain, *Radiology* **176**, 509–515, 1990.

20. H. Bruhn, G. Stoppe, K. D. Merboldt, T. Michaelis, W. Hänicke, and J. Frahm: Cerebral metabolic alterations in normal aging and Alzheimer's dementia detected by proton magnetic resonance spectroscopy, *Proceedings, Twelfth Annual Meeting, Society of Magnetic Resonance in Medicine*, Berlin, 1992, p. 752.

21. P. Christiansen, P. Tofts, H. B. W. Larsson, M. Stubgaard, and O. Henriksen: The concentration of N-acetyl aspartate, creatine/phosphocreatine and choline in different parts of the brain in adulthood and senium, *Proceedings, Twelfth Annual Meeting, Society of Magnetic Resonance in Medicine*, Berlin, 1992, p. 1932.

22. H. C. Charles, F. Lazeyras, K. R. Krishnan, O. B. Boyko, L. J. Patterson, P. M. Doraiswamy, and W. M. McDonald: Proton spectroscopy of human brain, effect of age and sex, *Biol. Psychiatry* **18(6)**, 995–1004, 1994.

23. H. J. Wittsack, H. Kugel, B. Roth, and W. Heindel: Quantitative measurements with localized ¹H MR spectroscopy in children with Canavan's disease, *J. Magn. Reson. Imag.* **6(6)** 889–893, 1996.

24. S. J. Austin, A. Connelly, D. G. Gadian, J. S. Benton, and E. M. Brett: Localized ¹H NMR spectroscopy in Canavan's disease: a report of two cases, *Magn. Reson. Med.* **19**, 439–445, 1991.

25. W. Grodd, I. Krägeloh-Mann, D. Peterson, D. F. K. Trefz, and K. Harzer: In vivo assessment of N-acetylaspartate in brain in spongy degeneration (Canavan's disease) by proton spectroscopy, *Lancet* **336,** 437–438, 1990.

26. C. C. Hanstock, D. L. Rothman, J. W. Prichard, T. Jue, and R. G. Shulman: Spatially localized ¹H-NMR spectra of metabolites in the human brain. *Proc. Natl. Acad. Sci. USA* **185**, 1821–1825, 1988.

27. R. Kreis, E. Arcinue, T. Ernst, T. K. Shonk, R. Flores, and B. D. Ross: Hypoxic encephalopathy after near-drowning studied by quantitative ¹H-magnetic resonance spectroscopy. *J. Clin. Invest.* **97(5)**, 1142–1154, 1996.

28. P. Gideon and O. Henricksen: In vivo relaxation of N-Acetyl-Aspartate, creatine plus phosphocreatine, and choline containing compounds during the

course of brain infarction, a proton MRS study, *Magn. Reson. Imag.* **10,** 983–988, 1992.

29. P. Gideon, O. Henriksen, B. Sperling, P. Christiansen, T. S. Olsen, H. S. Jorgensen, and P. Arlien-Soborg: Early time course of N-acetylaspartate, creatine and phosphocretine, and compounds containing choline in the brain after acute stroke. A proton magnetic resonance spectroscopy study, *Stroke* **23(11)**, 1566–1572, 1992.

30. V. P. Mathews, P. B. Barker, S. J. Blackband, J. C. Chatham, and R. N. Bryan: Cerebaral metabolites in patients with acute and subacute strokes: concentrations determined by quantitative proton MR spectroscopy, *AJR* **165**, 633–638, 1995.

31. W. Negengank: Studies of human tumors by MRS: a review, *NMR Biomed.* **5(5)** 303–324, 1992.

32. D. J. Meyerhoff, R. D. S. Mackay, N. Grossman, C. Van Dyke, G. Fein, and M. W. Weiner: Effects of normal aging and Alzheimer's disease on cerebral ¹H metabolites, *Proceedings, Twelfth Annual Meeting, Society of Magnetic Resonance in Medicine*, Berlin, 1992, p. 1931.

33. J. M. Constans, D. J. Meyerhoff, J. Gerson, S. Mackay, D. Norman, G. Fein, and M. W. Weiner: ¹H MR spectroscopic imaging of white matter hyperintensities: Alzheimer disease and ischemic vascular dementia, *Radiology* **197** 517–523, 1995.

34. S. Mackay, F. Ezekel, V. Di Sclafani, D. J. Meyerhoff, J. Gerson, D. Norman, G. Fein, and M. W. Weiner: Alzheimer disease and subcortical vascular dementia: evaluation by combining MR imaging segmentation and ¹H MR spectroscopic imaging, *Radiology* **198**, 537–545, 1996.

35. A. Connely, G. D. Jackson, J. S. Duncan, M. D. King, and D. G. Gadian: Magnetic resonance spectroscopy in temporal lobe epilepsy, *Neurology* **44**, 1411–1417, 1994.

36. H. Bruhn, J. Frahm, M. L. Gyngell, K. D. Merboldt, W. Hänicke, and R. Sauter: Cerebral metabolism in man after acute stroke: new observations using localized proton NMR spectroscopy, *Magn. Reson. Med* **9**, 126–131, 1989.

37. O. Henriksen, P. Gideon, B. Sperling, T. S. Olsen, H. S. Jorgensen, and P. Arlien-Soborg: Cerebral lactate production and blood flow in acute stroke, *J. Magn. Reson. Imag.* **2**, 511–517, 1992.

38. G. D. Graham, A. M. Blamire, D. L. Rothman, L. M. Brass, P. B. Fayad, O. A. C. Petroff, and J. W. Prichard: Early temporal variation of cerebral metabolites after human stroke: a proton magnetic resonance spectroscopy study, *Stroke* **24**, 1891–1896, 1993.

39. J. H. Cross, D. G. Gadian, A. Connely, and J. V. Leonard: Proton magnetic resonance spectroscopy in lactic acidosis and mitochondrial disorders, *J. Inher. Metab. Dis.* **16**, 800–811, 1993.

40. P. M. Matthews, F. Andermann, K. Silver, G. Karpati, and D. L. Arnold: Proton MR spectroscopic characterization of differences in regional brain

metabolic abnormalities in mitochondrial encephalomyopathies, *Neurology* **43**, 2484–2490, 1993.

41. M. Castillo, L.Kwock, and C. Green: MELAS Syndrome: Imaging and proton MR spectroscopic findings, *AJNR* **16**, 233–239, 1995.

42. S. N. Breiter, P. B. Barker, V. P. Mathews, S. S. Arroyo, and R. N. Bryan: Proton magnetic resonance spectroscopy in patients with seizure disorders, *Proceedings, Twelfth Annual Meeting, Society of Magnetic Resonance in Medicine*, Berlin, 1992, p. 644.

43. J. Frahm, G. Krüger, K. D. Merboldt, and A. Kleinschmidt: Dynamic uncoupling and recoupling of perfusion and oxidative metabolism during focal brain activation in man, *Magn. Reson. Med.* **35**, 143–148, 1996.

44. J. Prichard, D. Rothman, E. Novotny, O. Petroff, T. Kuwabara, M. Avison, A. Howseman, C. Hanstock, and R. Shulman: Lactate rise detected by ^{1}H NMR in human visual cortex during physiologic stimulation, *Proc. Natl. Acad. Sci. USA* **88**, 5829–5831, 1991.

45. F. D. Jungling, A. K. Wakhloo, and J. Hennig: In vivo proton spectroscopy of meningioma after preoperative embolization, *Magn. Reson. Med.* **30**, 155–160, 1993.

46. S. S. Gill, D. G. T. Thomas, N. Van Bruggen, D. G. Gadian, C. J. Peden, J. D. Bell, I. J. Cox, D. K. Menon, R. A. Iles, D. J. Bryant, and G. A. Coutts: Proton MR spectroscopy of intracranial tumours, in vivo and in vitro studies, *J. Comp. Assis. Tomog.* **14(4)**, 497–504, 1990.

47. B. L. Miller, R. A. Moats, T. Shonk, T. Ernst, S. Woolley, and B. D. Ross: Alzheimer disease: depiction of increased cerebral myo-inositol with proton MR spectroscopy, *Radiology* **187**, 433–437, 1993.

48. T. K. Shonk, R. A. Moats, P. Gifford, T. Michaelis, J. C. Mandigo, J. Izumi, and B. D. Ross: Probable Alzheimer disease: diagnosis with proton MR spectroscopy, *Radiology* **195**, 65–72, 1995.

49. R. Kreis and B. D. Ross: Cerebral metabolic disturbances in patients with subacute and chronic diabetes mellitus with proton magnetic resonance spectroscopy, *Radiology* **184**, 123–130, 1992.

50. R. H. Mattson, O. Petroff, D. Rothman, and K. Behar: Vigabatrin: effects on human brain GABA levels by nuclear magnetic resonance spectroscopy, *Epilepsia* **35** (Suppl. 5), S29–S32, 1994.

51. O. A. Petroff, K. L. Behar, R. H. Mattson, and D. L. Rothman: Human brain gamma-aminobutyric acid levels and seizure control following initiation of vigabatrin therapy, *J. Neurochem.* **67(6)**, 2399–2404, 1996.

52. O. A. Petroff, D. L. Rothman, K. L. Behar, T. L. Collins, and R. H. Mattson: Human brain GABA levels rise rapidly after initiation of vigabatrin therapy, *Neurology* **47(6)**, 1567–1571, 1996.

53. R. Kreis, T. Ernst, and B. D. Ross: Development of the human brain, in vivo

quantification of metabolite and water content with proton magnetic resonance spectrocopy, *Magn. Reson. Med.* **30**, 424–437, 1993.

54. H. Kimura, Y. Fujii, S. Itoh, T. Matsuda, T. Iwasaki, M. Maeda, Y. Konishi, and Y. Ishii: Metabolic alterations in the neonate and infant brain during development, evaluation with proton MR spectroscopy, *Radiology* **194**, 483–489, 1995.

55. C. J. Peden, F. M. Cowan, D. J. Bryant, J. Sargentoni, I. J. Cox, D. K. Menon, D. G. Gadian, J. D. Bell, and L. M. Dubowitz: Proton MR spectroscopy of the brain in infants, *J. Computed Assist. Tomogr.* **14(6)**, 886–894, 1990.

56. B. A. Holshouser, S. Ashwal, B. Y. Luh, S. Shu, S. Kahlon, K. L. Auld, L. G. Tomasi, R. M. Perkin, and D. B. Hinshaw: Proton MR spectroscopy after acute central nervous system injury: outcome prediction in neonates, infants, and children, *Radiology* **202**, 487–496, 1997.

57. G. Tedeschi, A. Righini, A. Bizzi, A. S. Barnett, and J. R. Alger: Cerebral white matter in the centrum semiovale exhibits a larger N-acetyl signal than does gray matter in long echo time ¹H-magnetic resonance spectroscopic imaging, *Magn. Reson. Med.* **33**, 127–133, 1995.

58. F. Lazeyras and H. C. Charles: Metabolic heterogeneity in normal brain tissue, *Proceedings, Tenth Annual Scientific Meeting, Society of Magnetic Resonance in Medicine*, San Francisco, 1991, p. 1061.

59. N. Schuff and M. W. Weiner: Investigation of metabolite changes in cortex and neocortex of healthy elderly by multislice ¹H MR spectroscopic imaging, *Proceedings, Fourth Scientific Meeting, International Society for Magnetic Resonance in Medicine*, New York, 1996, p. 1213.

60. R. Sauter: Cerebral single volume proton spectroscopy on healthy volunteers: a multicenter pilot study, *Proceedings, 12th Annual Scientific Meeting, Society for Magnetic Resonance in Medicine*, New York, 1993, p. 1531.

61. I. D. Wilkinson, K. A. Miszkiel, R. J. S. Chinn, M. Paley, M. A. Maloney, M. A. Hall-Craggs, and M. I. G. Harrison: The influence of gender on cranial ¹H MRS and volumetry, *Proceedings, Fifth Annual Meeting, International Society for Magnetic Resonance in Medicine*, Vancouver, 1997, p. 1163.

62. H. Bruhn, J. Frahm, M. L. Gyngell, K. D. Merboldt, W. Hänicke, R. Sauter, and C. Hamburger: Noninvasive differentiation of tumors with use of Localized ¹H MR spectroscopy in vivo: initial experience in patients with cerebral tumors, *Radiology* **172**, 541–548, 1989.

63. W. Negendank, R. Sauter, T. R. Brown, J. L. Evelhoch, A. Falini, E. D. Gotsis, A. Heerschap, K. Kamada, B. C. P. Lee, M. M. Mengeot, E. Moser, K. A. Padavic-Shaller, J. A. Sanders, T. A. Spraggins, A. E. Stillman, B. Terwey, T. J. Vogl, K. Wicklow, and R. A. Zimmerman: Proton magnetic resonance spectroscopy in patients with glial tumors: a multicenter study, *J. Neurosurg.* **84**, 449–458, 1996.

64. W. Negendank and R. Sauter: Intratumoral lipids in ¹H MRS in vivo in brain

tumors: experience of the Siemens cooperative clinical trial, *Anticancer Res.* **16(3B)**, 1533–1538, 1996.

65. J. S. Taylor, J. W. Langston, W. E. Reddick, P. B. Kingsley, R. J. Ogg, M. H. Pui, L. E. Kun, J. J. Jenkens III, G. Chen, J. J. Ochs, R. A. Sanford, and R. L. Heideman: Clinical value of proton magnetic resonance spectroscopy for differentiating recurrent or residual brain tumor from delayed cerebral necrosis, *Int. J. Rad. Oncol. Biol. Phys.* **36(5)**, 1251–1261, 1996.

66. K. Kamada, K. Houkin, H. Abe, Y. Sawamura, and T. Kashiwaba: Differentiation of cerebral radiation necrosis from tumor recurrence by proton magnetic resonance spectroscopy, *Neuro. Med. Chir. (Tokyo)* **37**, 250–256, 1997.

67. L. Chang, B. L. Miller, D. McBride, M. Cornford, G. Oropilla, S. Buchtal, F. Chiang, and T. Ernst: Brain lesions in patients with aids: ^{1}H MR Spectroscopy, *Radiology* **197**, 525–531, 1995.

68. P. B. Barker, R. R. Lee, and J. C. McArthur: AIDS dementia complex: evaluation with proton MR spectroscopic imaging, *Radiology* **195**, 58–64, 1995.

69. W. K. Chong, M. Paley, I. D. Wilkinson, M. A. Hall-Craggs, B. Sweeney, M. J. G. Harrison, R. F. Miller, and B. E. Kendall: Localized cerebral proton MR spectroscopy in HIV infection and AIDS, *AJNR* **15**, 21–25, 1994.

70. D. J. Meyerhoff, S. Mackay, N. Poole, W. P. Dillon, M. W. Weiner, and G. Fein: N-acetylaspartate reductions measured by ^{1}H MRSI in cognitively impaired HIV-seropositive individuals, *Magn. Reson. Imag.* **12**, 653–659, 1994.

71. D. J. Meyerhoff, M. W. Weiner, and G. Fein: Deep gray matter structures in HIV infection: a proton MR spectroscopic study, *AJNR* **17**, 973–978, 1966.

72. J. Wolinsky, P. A. Narayana, and M. J. Fenstermacher: Proton magnetic resonance spectroscopy in multiple sclerosis, *Neurology* **40**, 1764–1769, 1990.

73. C. A. Davie, C. P. Hawkins, G. J. Barker, A. Brennan, P. S. Tofts, D. H. Miller, and W. I. MacDonald: Serial proton magnetic resonance spectroscopy in acute multiple sclerosis lesions, *Brain* **117**, 49–58, 1994.

74. S. E. Davies, J. Newcombe, S. R. Williams, W. I. MacDonald, and J. B. Clark: High resolution proton NMR spectroscopy of multiple sclerosis lesions, *J. Neurochem.* **64**, 742–748, 1995.

75. W. Roser, G. Hagberg, I. Mader, H. Brunnschweiler, E. W. Radue, J. Seelig, and L. Kappos: Proton MRS of gadolinium-enhancing MS plaques and metabolic changes in normal-appearing white matter, *Magn. Reson. Med.* **33**, 811–817, 1995.

76. I. L. Simone, F. Federico, M. Trojano, C. Tortorella, M. Liguori, P. Giannini, E. Picciola, G. Natile, and P. Livrea: High resolution proton MR spectroscopy of cerebrospinal fluid in MS patients. Comparison with biochemical changes in demyelinating plaques, *J. Neurol. Sci.* **144**, 182–190, 1996.

77. G. Ende, K. D. Laxer, R. C. Knowlton, G. B. Matson, N. Schuff, G. Fein, and M. W. Weiner: Temporal lobe epilepsy: bilateral hippocampal metabolite changes revealed at proton MR spectroscopic imaging, *Radiology* **202**, 809–817, 1997.

78. S. H. Kim, K. H. Chang, I. C. Song, M. H. Han, H. C. Kim, H. S. Kang,and M. C. Han: Brain abcess and brain tumor: discrimination with in vivo ^{1}H spectroscopy, *Radiology*, **204**, 239–245, 1997.

79. R. Canese, G. S. Payne, and M. O. Leach: A quantitative analysis of the effect of Gd-DTPA contrast agents on ^{1}H MR spectra of brain tumours, *Proceedings, Fourth Scientific Meeting, International Society for Magnetic Resonance in Medicine*, New York, 1996, p. 1196.

80. J. S. Taylor, W. E. Reddick, P. B. Kingsley, and R. J. Ogg: Proton MRS after Gadolinium contrast agent, *Proceedings, Third Scientific Meeting and Exhibition, International Society for Magnetic Resonance in Medicine*, Nice, 1995, p. 1854.

81. E. T. Kiriakopoulos, C. A. Stewart, B. M. Guthrie, R. Walcarius, and D. J. Mikulis: The effect of Gd-DTPA on ^{1}H proton spectroscopy in the normal human brain, *Proceedings, Second Scientific Meeting, International Society for Magnetic Resonance in Medicine*, San Francisco, 1994, p. 566.

82. P. E. Sijens, M. J. Van Den Bent, P. J. C. M. Nowak, P. van Dijk, and M. Oudkerk: [1]H chemical shift imaging reveals loss of brain tumor choline signal after administration of a Gd-contrast, *Magn. Reson. Med.* **37**, 222–225, 1997.

83. M. Barany, P. N. Venkatasubramanian, E. Mok, I. M. Siegel, E. Abraham, N. D. Wycliffe, and M. F. Mafee: Quantitative and qualitative fat analysis in human leg muscle of neuromuscular diseases by ^{1}H MR spectroscopy in vivo, *Magn. Reson. Med.* **10**, 210–226, 1989.

84. H. Bruhn, J. Frahm, M. L. Gyngell, K. D. Merboldt, W. Hänicke, and R. Sauter: Localized proton NMR spectroscopy using stimulated echoes, applications to human skeletal muscle in vivo, *Magn. Reson. Med.* **17**, 82–94, 1991.

85. F. Schick, B. Eismann, W. I. Jung, H. Bongers, M. Bunse, and O. Lutz: Comparison of localized proton NMR signals of skeletal muscle and fat in vivo, two lipids compartments in muscle tissue, *Magn. Reson. Med.* **29**, 158–167, 1993.

86. P. A. Bottomley, Y. H. Lee, R. G. Weiss: Total creatine in muscle, Imaging and quantification with proton MR spectroscopy, *Radiology* **204**, 403–410, 1997.

87. F. Schick, H. Bongers, W. I. Jung, M. Skalej, O. Lutz, and C. D. Claussen: Volume selective proton MRS in vertebral bodies, *Magn. Reson. Med.* **26**, 207–217, 1992.

88. K. E. Jensen, M. Jensen, P. Grundtvig, C. Thomsen, H. Karle, and O. Henriksen: Localized in vivo proton spectroscopy of the bone marrow in patients with leukemia, *Magn. Reson. Imag.* **8**, 779–789, 1990.

89. D. Ballon, A. Jakubowski, J. Gabrilove, M. C. Graham, M. Zakowski, C. Sheridan, and J. A. Koutcher: In vivo measurements of bone marrow cellularity using volume-localized proton NMR spectroscopy, *Magn. Reson. Med.* **19**, 85–95, 1991.

90. J. R. Roebuck, R. E. Lenkinski, L. Bolinger, C. P. Langlotz, and M. D. Schnall: Proton magnetic resonance spectroscopy of human breast disease, *Syllabus,*

MR of Cancer: Physiology and Metabolism, International Society for Magnetic Resonance in Medicine, John Hopkins University, Baltimore, August 8 and 9, 1996, p. 35.

91. J. Kurhanewicz, D. B. Vigneron, S. J. Nelson, H. Hricack, S. Moyher, P. Carrol, T. L. James, and P. Narayan: In vivo citrate levels in the normal and pathologic human prostate, *Proceedings, Twelfth Annual Scientific Meeting, Society for Magnetic Resonance in Medicine*, New York, 1993, p. 212.

92. J. Kurhanewicz, S. Nelson, S. Moyher, L. Carvajal, H. Hricack, P. Carrol, and D. Vigneron: Following metabolic response to prostate cancer hormone therapy using three dimensional ^1H spectroscopic imaging, *Proceedings, Fourth Scientific Meeting, International Society for Magnetic Resonance in Medicine*, New York, 1996, p. 1053.

93. D. B. Vigneron, J. Kurhanewicz, and S. Nelson: Prostate magnetic resonance spectroscopy, *Syllabus, MR of Cancer: Physiology and Metabolism*, International Society for Magnetic Resonance in Medicine, John Hopkins University, Baltimore, August 8 and 9, 1996, p. 11.

94. J. Kurhanewicz, D. Vigneron, H. Hricack, P. Narayan, P. Carrol, and S. Nelson: Three-dimensional ^1H MR spectroscopic imaging of the in situ human prostate with high (0.24–0.7 cm3) spatial resolution, *Radiology* **198**, 795–805, 1996.

95. A. Heerschap, G. J. Jager, M. Van der Graaf, J. O. Barentsz, J. J. de la Rosette, G. O. Oosterhof, E. T. Ruijter, and S. J. Ruijs: In vivo proton MR spectroscopy reveals altered metabolites content in malignant prostate tissue, *Anticancer Res.* **17(3A)**, 1455–1460, 1997.

96. R. Longo, R. Ricci, F. Masutti, R. Vidimari, L. S. Croce, L. Bercich, C. Tiribelli, and L. Dalla Palma: Fatty infiltration of the liver, quantification by ^1H localized magnetic resonance spectroscopy and comparison with computed tomography, *Invest. Radiol.* **28**, 297–302, 1993.

97. O. A. C. Petroff, J. W. Prichard, K. L. Behar, J. R. Alger, J. A. den Hollander, and R. G. Shulman: Cerebral intracellular pH by ^{31}P nuclear magnetic resonance spectroscopy, *Neurology* **35**, 781–786, 1985.

98. C. Boesch, R. Gruetter, E. Martin, G. Duc, and K. Wüthrich: Variations in the in vivo ^{31}P MR spectra of the developing human brain during postnatal life, *Radiology* **172**, 197–199, 1989.

99. D. Azzopardi, J. S. Wyatt, P. A. Hamilton, E. B. Cady, D. T. Delpy, P. L. Hope, and E. O. R. Reynolds: Phosphorus metabolites and intracellular pH in the brains of normal and small for gestational age infants investigated by magnetic resonance spectroscopy, *Ped. Res.* **25(5)**, 440–444, 1989.

100. T. Higuchi, C. Tanaka, S. Naruse, and T. Ebisu: T2 weighted in vivo ^{31}P MRS in human brain, *Proceedings, Ninth Annual Scientific Meeting, Society of Magnetic Resonance in Medicine*, New York, 1990, p. 992.

101. C. Tanaka, T. Higuchi, S. Naruse et al.: ^{31}P MRS associated with development of human brain, *Proceedings, Eighth Annual Scientific Meeting, Society of Magnetic Resonance in Medicine*, Amsterdam, 1989, p. 462.

102. S. R. Levine, J. A. Helpern, K. M. Welsh, A. M. Vande Linde, K. L. Sawaya, E. E. Brown, N. M. Ramadan, R. K. Deveshwar, and R. J. Ordidge: Human

focal cerebral ischemia: evaluation of brain pH and energy metabolism with ^{31}P NMR spectroscopy, *Radiology* **185**, 537–544, 1992.

103. O. A. Petroff, G. D. Graham, A. M. Blamire, M. al-Rayess, D. L. Rothman, P. B. Fayad, L. M. Brass, R. G. Shulman, and J. W. Prichard: Spectroscopic imaging of stroke in humans, histopathology correlates of spectral changes, *Neurology* **42**, 1349–1354, 1992.

104. D. Sappey-Marinier, B. Hubesch, G. B. Matson, and M. W. Weiner: Decreased phosphorus metabolite concentrations and Alkalosis in chronic cerebral infarction, *Radiology* **182**, 29–34, 1992.

105. Y. Nagai, S. Naruse, and M. W. Weiner: Effect of hypoglycemia on changes of brain lactic acid and intracellular pH produced by ischemia, *NMR Biomed.* **6**, 1–6, 1993.

106. D. L. Arnold, J. F. Emrich, E. A. Shoubridge, J. G. Villemure, and W. Feindel: Characterization of astrocytomas, meningiomas, and pituitary adenomas by phosphorus magnetic resonance spectroscopy, *J. Neurosurg.* **74**, 447–453, 1991.

107. B. Hubesch, D. Sappey-Marinier, K. Roth, D. J. Meyerhoff, G. B. Matson, and M. W. Weiner: ^{31}P MR spectroscopy of normal human brain and brain tumors, *Radiology* **174**, 401–409, 1990.

108. P. A. Bottomley, C. J. Hardy, J. P. Cousins, M. Armstrong, and W. A. Wagle: AIDS dementia complex, brain high-energy phosphaste metabolic deficits, *Radiology* **176**, 407–411, 1990.

109. R. F. Deicken, B. Hubesch, P. C. Jensen, D. Sappey-Marinier, P. Krell, A. Wisniewski, D. Vanderburg, R. Parks, G. Fein, and M. W. Weiner: Alteration in brain phosphate metabolite concentrations in patients with human immunodeficiency virus infection, *Arch. Neurol.* **48**, 203–209, 1991.

110. J. M. Minderhound, E. L. Mooyaart, R. L. Kamman, A. W. Teelken, M. C. Hoogstraten, L. M. Vencken, E. Gravenmade Jr., and W. Van den Burg: In vivo phosphorus magnetic resonance spectroscopy in multiple sclerosis, *Arch. Neurol.* **49**, 161–165, 1992.

111. C. A. Husted, D. S. Goodin, J. W. Hugg, A. A. Maudsley, J. S. Tsuruda, S. H. de Bie, G. Fein, G. B. Matson, and M. W. Weiner: Biochemical alterations in multiple sclerosis lesions and normal-appearing white matter detected by in vivo ^{31}P and ^{1}H spectroscopic imaging, *Ann. Neurol.* **36(2)**, 157–165, 1994.

112. P. A. Bottomley, J. P. Cousins, D. J. Hardy, D. L. Pendry, W. A. Wagle, C. J. Hardy, F. A. Earnes, R. J. McCafrey, and D. A. Thomson: Alzheimer dementia: quantification of energy metabolism and mobile phosphoesters with ^{31}P NMR spectroscopy, *Radiology* **183**, 696–699, 1992.

113. G. G. Brown, S. R. Levine, J. M. Gorrel, J. W. Pettegrew, J. W. Gdowski, J. A. Bueri, J. A. Helpern, and K. M. A. Welch: In vivo ^{31}P NMR profiles of Alzheimer's disease and multiple subcortical infarct dementia, *Neurology* **39**, 1423–1427, 1989.

114. J. W. Pettegrew, G. Withers, K. Panchalingram, and J. F. M. Post: ^{31}P nuclear magnetic resonance (NMR) spectroscopy of brain in aging and Alzheimer's disease, *Neural. Transm.* **24** (Supp. 1), 261–268, 1987.

115. C. D. Smith, L. G. Gallenstein, W. J. Layton, R. J. Kryscios, and W. R. Markesbery: [31]P magnetic resonance spectrosopy in Alzheimer's and Pick's disease, *Neurobiol. Aging* **14**, 85–92, 1993.

116. J. W. Pettegrew, W. E. Klunk, E. Kanal, K. Panchalingam, and R. J. McClure: Changes in brain membrane phospholipid and high-energy phosphate metabolism precede dementia, *Neurobiol. Aging* **16**, 973–975, 1995.

117. W. E. Klunk, K. Panchalingam, R. J. McClure, and J. W. Pettegrew: Quantitative ^{1}H and ^{31}P MRS of PCA extraxts postmortem Alzheimer's disease brain, *Neurobiol. Aging* **17**, 349–357, 1996.

118. R. G. Gonzales, A. R. Guimaraes, G. J. Moore, A. Crawley, L. A. Cupples, and J. H. Growdon: Quantitative in vivo ^{31}P magnetic resonance spectroscopy of Alzheimer disease, *Alzheimer Dis. Assoc. Disord.* **10**, 46–52, 1996.

119. J. W. Hugg, K. D. Laxer, G. B. Matson, A. A. Maudsley, C. A. Husted, and M. W. Weiner: Lateralization of human focal epilepsy by ^{31}P magnetic resonance spectroscopic imaging, *Neurology* **42**, 2011–2018, 1992.

120. R. Kuzniecky, G. A. Elgavish, H. P. Hetherington, W. T. Evanochko, and G. M. Pohost: In vivo ^{31}P nuclear magnetic resonance spectroscopy of human temporal lobe epilepsy, *Neurology* **42,** 1586–1590, 1992.

121. K. D. Laxer, B. Hubesch, D. Sappey-Marinier, and M. W. Weiner: Inceased pH and inorganic phosphate in temporal seizure foci demonstrated by ^{31}P MRS, *Epilepsia* **33(4)**, 618–623, 1992.

122. J. W. Hugg, G. B. Matson, D. B. Twieg, A. A. Maudsley, D. Sappey-Marinier, and M. W. Weiner: Phosphorus-31 MR spectroscopic imaging (MRSI) of normal and pathological human brains, *Magn. Reson. Imag.* **10(2)**, 227–243, 1992.

123. P. A. Garcia, K. D. Laxer, and T. Ng: Application of spectroscopic imaging in epilepsy, *Magn. Reson. Imag.* **13(8)**, 1181–1185, 1995.

124. B. Barbiroli, P. Montagna, P. Cortelli, R. Funicello, S. Iotti, L. Monari, G. Pierangeli, P. Zaniol, and E. Lugaresi: Abnormal brain and muscle energy metabolism shown by ^{31}P magnetic resonance spectroscopy in patients affected by migraine with aura, *Neurology* **42**, 1209–1214, 1992.

125. P. Montagna, P. Cortelli, L. Monari, G. Pierangeli, P. Parchi, R. Lodi, S. Iotti, C. Frassineti, P. Zaniol, E. Lugaresi, et al.: [31]P magnetic resonance spectroscopy in migraine without aura, *Neurology* **44(4)**, 666–669, 1994.

126. K. M. Welch, S. R. Levine, G. D'Andrea, L. R. Schultz, and J. A. Helpern: Preliminary observations on brain energy metabolism in migraine studied by in vivo phosphorus ^{31}P NMR spectroscopy, *Neurology* **39**, 538–541, 1989.

127. P. Montagna: Magnetic resonance spectroscopy in migraine: *Cephalalgia* **15(4)**, 323–327, 1995.

128. G. J. Kemp and G. K. Radda: Quantitative interpretation of bioenergetic data from ^{31}P and ^{1}H magnetic resonance spectroscopic studies of skeletal muscle: an analytical review, *Magn. Reson. Q.* **10(1)**, 43–63, 1994.

129. B. Chance, J. S. Leigh, B. J. Clark, J. Maris, J. Kent, S. Nioka, and D. Smith: Control of oxidative metabolism and oxygen delivery in human skeletal

muscle: a steady state analysis of the work energy cost transfer function, *Proc. Natl. Acad. Sci. USA* **82**, 8384–8388, 1985.

130. D. Bendahan, S. Confort-Gouny, G. Kozak-Reiss, and P. J. Cozzone: Heterogeneity of metabolic response to muscular exercise in humans: New criteria of invariance defined by in vivo phosphorus-31 NMR spectroscopy, *FEBS Lett.* **272**, 155–158, 1990.

131. D. L. Arnold, D. J. Taylor, and G. K. Radda: Investigation of human mitochondrial myopathies by phosphorus magnetic resonance spectroscopy, *Ann. Neurol.* **18**, 189–196, 1985.

132. Z. Argov, W. J. Bank, J. Maris, P. Peterson, and B. Chance: Bioenergetic heterogeneity of human mitochondrial myopathies, phosphorus magnetic resonance spectroscopy study, *Neurology* **37**, 257–262, 1987.

133. J. H. Park, J. P. Vansant, N. G. Kumar, S. J. Gibbs, M. S. Curvin, R. R. Price, C. L. Partain, and A. E. James Jr: Dermatomyositis, correlative MR imaging and ^{31}P MR spectroscopy for quantitative characterization of inflammatory disease, *Radiology* **177**, 473–479, 1990.

134. R. Zahler, S. Majumdar, B. Frederick, M. Laughlin, E. Barrett, and J. C. Gore: NMR determination of myocardial pH in vivo, separation of tissue inorganic phosphate from blood 2,3-DPG, *Magn. Reson. Med.* **17**, 368–378, 1991.

135. J. R. Gober, G. G. Schwartz, S. Schaefer, B. M. Massie, G. B. Matson, M. W. Weiner, and G. S. Karczmar: ^{31}P MRS of myocardial inorganic phosphate using radiofrequency gradient echoes, *Magn. Reson. Med.* **20**, 171–183, 1991.

136. P. A. Bottomley: MR spectroscopy of the human heart: The status and the challenges, *Radiology* **191**, 593–612, 1994.

137. P. A. Bottomley: Human in vivo NMR spectroscopy in diagnostic medicine: Clinical tool or research probe? *Radiology* **170**, 1–15, 1989.

138. C. E. Haug, J. I. Shapiro, L. Chan, and R. Weil: ^{31}P nuclear magnetic resonance spectroscopic evaluation of heterotopic cardiac allograft rejection in the rat, *Transplantation* **44**, 175–178, 1987.

139. W. T. Evanochko, A. Bouchard, J. K. Kirklin, R. C. Bourge, D. Luney, and G. M. Prohost: Detection of Cardiac Transplant rejection in Patients by ^{31}P NMR Spectroscopy, *Proceedings, Ninth Annual meeting, Society of Magnetic Resonance in Medicine*, New York, 1990, p. 246.

140. R. J. Herfkens, H. C. Charles, R. Negro-Velar, and P. Van Trigg: In vivo ^{31}P NMR spectroscopy of human heart transplants, *Proceedings, Seventh Annual Meeting, Society of Magnetic Resonance in Medicine*, San Francisco, 1988, p. 827.

141. S. Masson, O. Henrikson, A. Stengaard, C. Thomsen, and B. Quistorff: Hepatic metabolism during constant infusion of fructose: comparative studies with ^{31}P magnetic resonance spectroscopy in man and rats, *Biochim. Biophys. Acta* **1199**, 166–174, 1994.

142. R. Oberhaensli, B. Rajagopalan, G. J. Gallaway, D. J. Taylor, and G. K. Radda: Study of human liver disease with ^{31}P magnetic resonance spectroscopy, *Gut* **31**, 463–467, 1990.

143. D. J. Meyerhoff, G. S. Karczmar, and M. W. Weiner: Abnormalities of the liver evaluated by ^{31}P MRS, *Invest. Radiol.* **24**, 980–984, 1989.

144. D. J. Meyerhoff, G. S. Karczmar, F. Valone, A. Venook, G. B. Matson, and M. W. Weiner: Hepatic cancers and their response to chemoembolization therapy, quantitative image-guided ^{31}P magnetic spectroscopy, *Invest. Radiol.* **27**, 456–464, 1992.

145. D. J. Meyerhoff, M. D. Boska, A. M. Thomas, and M. W. Weiner: Alcoholic liver disease, quantitative image-guided ^{31}P MR spectroscopy, *Radiology* **173**, 393–400, 1989.

146. R. Kalra, K. E. Wade, L. Hands, P. Styles, R. Camplejohn, M. Greenall, G. E. Adams, A. L. Harris, and G. K. Radda: Phosphomonoester is associated with proliferation in human breast cancer: a ^{31}P MRS study, *Br. J. Cancer* **67**, 1145–53, 1993.

147. P. E. Sijens, H. K. Wijrdeman, M. A. Moerland, C. J. G. Bakker, J. W. Vermeulen, and P. R. Luyten: Human Breast Cancer in vivo: ^{1}H and ^{31}P MR spectroscopy at 1.5T, *Radiology* **169**, 615–620, 1988.

148. C. J. Twelves, D. A. Porter, M. Lowry, N. A. Dobbs, P. E. Graves, M. A. Smith, R. D. Rubens, and M. A. Richards: Phosphorus-31 metabolism of post-menopausal breast cancer studied in vivo by magnetic resonance spectroscopy, *Br. J. Cancer* **69**(9), 1151–1156, 1994.

149. P. Jehenson, D. Duboc, G. Bolch, M. Fardeau, and A. Syrota: Diagnosis of muscular glycogenosis by in vivo natural abundance ^{13}C NMR spectroscopy, *Neuromusc. Disord.* **1**, 99–101, 1991.

150. G. I. Shulman, D. L. Rothman, T. Jue, P. Stein, R. A. DeFronzo, and R. G. Shulman: Quantitation of muscle glycogen synthesis in normal subjects and subjects with non-insulin-dependent diabetes by ^{13}C nuclear magnetic resonance spectroscopy, *N. Eng. J. Med.* **25**, 233–238, 1990.

151. W. Wolf, M. J. Albright, M. S. Silver, H. Weber, U. Reichardt, and R. Sauter: Fluorine-19 NMR spectroscopy studies of the metabolism of 5-fluorouracil in the liver of patients undergoing chemotherapy, *Magn. Reson. Imag.* **5**, 165–169, 1987.

152. C. A. Presant, W. Wolf, M. J. Albright, K. L. Servis, R. Ring, D. Atkinson, R. L. Ong, C. Wiseman, M. King, D. Blayney, et al.: Human tumor fluorouracil trapping, clinical correlations of in vivo ^{19}F nuclear magnetic resonance spectroscopy pharmacokinetics, *J. Clin. Oncol.* **8**, 1868–1873, 1990.

153. C. A. Presant, W. Wolf, V. Waluch, C. Wiseman, P. Kennedy, D. Blayney, and R. R. Brechner: Association of intratumoral pharmacokinetics of fluorouracil with clinical response, *Lancet* **343**, 1184–87, 1994.

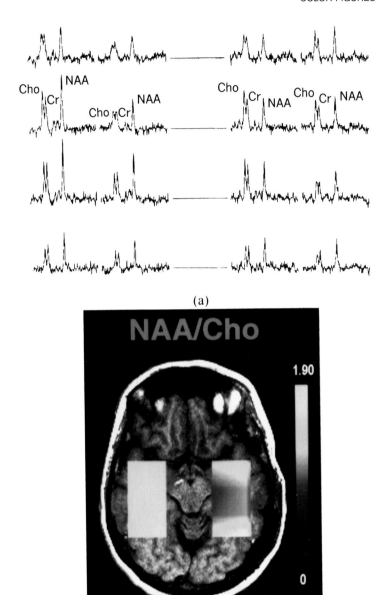

(a)

(b)

Figure 6.13. An illustration of a unilateral TLE scan. (*a*) An array plots of the 2 × 4 spectra of both temporal lobes. Spectral profile of the mesial area in the left temporal lobe has NAA/Cho < 1 while other voxels (including the contralateral TL) all show NAA/Cho > 1. (*b*) Metabolite NAA/Cho map in color overlaid on a gray scale T1-weighted image showing low signal intensity in the left anterior region. [Reproduced by permission of RSNA Publications from T. C. Ng et al., *Radiology, 193*, 465–472 (1994).]

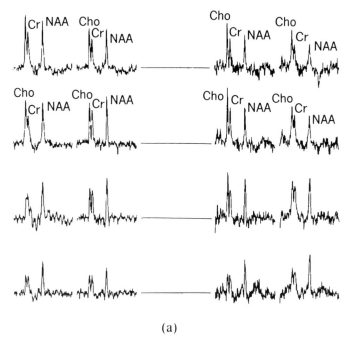

(a)

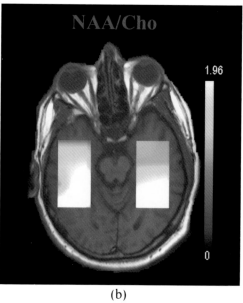

(b)

Figure 6.14. The CSI results of a patient suffering from bitemporal lobe epilepsy. (*a*) The array of the bitemporal lobe spectra illustrating that the abnormal NAA/Cho is more severe in the left anterior mesial (involves 5 voxels) than in the lateral right TL. (*b*) The overlay of color NAA/Cho map on a T1-weighted image also shows low signal in the same regions as shown in (*a*). [Reproduced by permission of RSNA Publications from T. C. Ng et al., *Radiology, 193,* 465–472 (1994).]

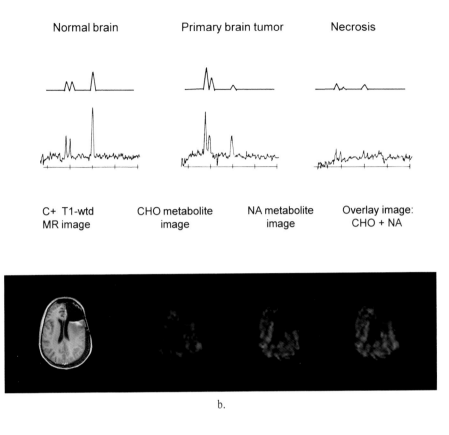

Normal brain   Primary brain tumor   Necrosis

C+ T1-wtd   CHO metabolite   NA metabolite   Overlay image:
MR image   image   image   CHO + NA

b.

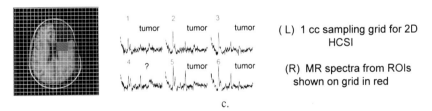

1   2   3
tumor   tumor   tumor

4   ?   5   tumor   6   tumor

(L) 1 cc sampling grid for 2D
    HCSI

(R) MR spectra from ROIs
    shown on grid in red

c.

Figure 6.19. Proton MR spectroscopy used in the differentiation of tumor from delayed radiation necrosis. (*a*) The MRS can distinguish between tumor (TU) and delayed radiation necrosis (DN): Prospective MRS identified 5/7 TU and 4/5 DN cases by characteristic metabolic patterns. (*b*) The 2D CSI for TU versus DN correctly predicts TU: In the contrast-enhancing regions, Cho is isointense to elevated (compared to normal white matter) and NA is decreased. (*c*) (*L*) The 1-cm^3 sampling grid for 2D HCSI. (*R*) MR spectra from ROIs shown on grid in red. [Courtesy of J. Taylor, St. Jude Children's Research Hospital, Memphis, TN.]

Early Localization Techniques

The current approaches for signal localization described in Chapter 4 are derived from imaging techniques and use standard imaging hardware. Other techniques that were proposed prior to the commercial development of MRI systems involved the use of surface coils as transmitters and/or receivers as well as spatial variation of the static magnetic field (B_0) or the rf field (B_1). While these methods have generally been replaced by the SVS and CSI techniques, they are included here to provide a historical perspective of the methodology.

A. ROTATING FRAME ZEUGMATOGRAPHY

Rotating frame zeugmatography[1,2] is a 1D technique that exploits the $\mathbf{B_1}$ field gradient characteristic of a transmitting surface coil. This gradient is such that the $\mathbf{B_1}$ field decreases with increasing distance from the coil. Therefore, those spins closest to the coil experience the largest flip angle. The same spins experience the most variation in flip angle when the duration of the rf pulse or the amplitude of $\mathbf{B_1}$ is changed (Fig. A.1). These properties may be utilized for spatial encoding along the axis of a surface coil. In rotating frame zeugmatography, MRS data (FID) are collected in a series of experiments where the rf pulse duration or the amplitude of $\mathbf{B_1}$ is varied. Two-dimensional FT is performed on the data to yield high-resolution spectra from various depths along the coil axis.

Rotating frame zeugmatography suffers from many drawbacks. There is signal contamination from regions close to the coil surface and from lateral tissue. Quantitation is difficult because the sensitivity for signal detection is dependent on the distance away from the coil. Three-dimensional localization requires additional transmit/receive coils and may be difficult to implement. There is a signal reduction on long duration pulses due to spin dephasing making short T2 metabolites hard to detect.

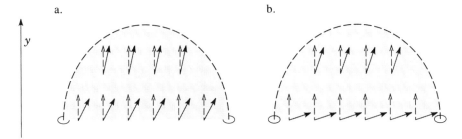

Figure A.1. (*a*) Flip angle decreases with increasing distance from coil surface. (*b*) Spins closer to coil surface experience larger variation in flip angle with increased pulse duration or pulse amplitude is increased.

B. DEPTH PULSES

The single volume depth pulse technique,[3,4] employs multiple rf pulses with phase cycling for cancellation of unwanted signals. The sequence starts with an rf pulse of flip angle α, followed by a waiting period τ, then a pulse 2α and another waiting period τ, after which the echo is collected. By varying the duration of pulse α, its flip angle is adjusted to be 90° over a region of interest. Phase of the 2α pulse is cycled such that it flips the magnetization about a different axis ($+x$, $-x$, $+y$, $-y$) in four successive scans. Signals collected from the $+x$ and $-x$ cycles are added and those from the $+y$ and $-y$ cycles are subtracted. Consequently, the final signal originates from an ROI that experiences a 90° flip angle while outside signals are canceled. The technique requires calibration of the α pulse before each measurement, which prolongs the examination time. Also, depth localization may not be accurate due to curvature in the rf field lines produced by a current loop. This problem is solved by using two decoupled coils, a large transmitting coil, and a smaller receiving coil. This, however, does not completely eliminate contamination from tissue away from the axis of the coil.

C. TOPICAL MAGNETIC RESONANCE

Topical magnetic resonance (TMR),[5,6] also called magnetic field profiling, is based on placing the ROI in a homogeneous $\mathbf{B_0}$ field and adjusting special gradients to spoil the field homogeneity elsewhere. A high-resolution spectrum is obtained from the homogeneous ROI. An underlying very broad hump comes from the external inhomogeneous regions and is subsequently removed during data processing. This technique eliminates signal contam-

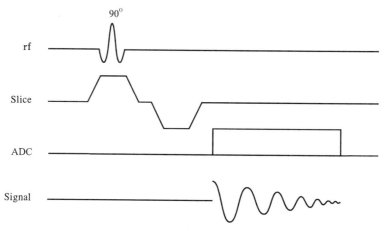

Figure A.2. Timing diagram of the DRESS sequence.

ination coming from tissues in the immediate proximity of the coil. However, defining the region from which the signal originates is difficult. In practice, the anatomy to be examined is placed at the center of the magnet. An additional field is produced by special coils affixed to the standard system configuration. The coils are designed so that field homogeneity is preserved only at the center of the magnet, which limits flexibility in patient positioning and requires repositioning when more than one region needs to be examined.

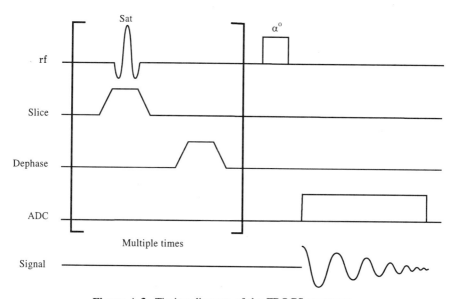

Figure A.3. Timing diagram of the FROGS sequence.

D. DEPTH-RESOLVED SURFACE COIL SPECTROSCOPY

A simple and well-known method for 1D surface coil localization is the *d*epth-*r*esolved *s*urface coil *s*pectroscopy or DRESS[7] sequence (Fig. A.2). It consists of a frequency selective 90° rf pulse applied simultaneously with a gradient perpendicular to the coil surface. A large transmitting surface coil is used for more uniform excitation of the plane of interest. A smaller receiver coil detects the signal originating from a slice smaller than the excited slice. Data collection is done right after the slice selection gradient is switched off. A multislice version of DRESS called slice-interleaved depth-resolved surface coil spectroscopy (SLIT DRESS)[8] allows acquisition of spectra from several slices in the same time it takes to acquire a single spectrum with the original DRESS sequence.

DRESS brings considerable improvement to surface coil localization. It achieves accurate spectral allocation to slices of tissue parallel to the surface of a coil with no significant contamination from outside tissue. However, it still has some limitations. For example, it does not allow spatial selectivity within a slice that may contain a variety of tissues such as a slice through the heart containing blood, myocardium, and epicardium. Also, the delay between the rf pulse and data sampling causes baseline distortions in the spectrum that may be difficult to eliminate with phase correction during postprocessing.

E. FAST ROTATING GRADIENT SPECTROSCOPY

Fast rotating gradient spectroscopy (FROGS) (Fig. A.3) is another single volume technique first proposed in 1987.[9] FROGS relies on selective spatial presaturation of tissue near the surface of the coil to remove unwanted signals. Presaturation is achieved by applying a frequency selective rf pulse simultaneously with a slice selective gradient perpendicular to the coil surface followed by a spoiler gradient. Efficiency of this process is increased by repeating it many times while varying the amplitude of the presaturation pulse and the duration and orientation of the dephasing gradient. Following presaturation, a nonselective α pulse produces a signal only from the part of the sensitive volume that was not included in the saturation region. The FROGS sequence can be modified for a different orientation of the saturation region or to include multiple saturation bands in different orientations to isolate the desired volume.

TABLE A.1. Early Localization Techniques for Surface Coils

Technique	Method	Description: Signal Detection
Rotating frame zeugmatography	$\mathbf{B}_1$ gradient	1D technique. It measures spectra from various depth along the axis of surface coil
Depth pulse	$\mathbf{B}_1$ gradient	It detects signal from a large VOI
Topical magnetic field (TMR)	$\mathbf{B}_0$ inhomogeneities	It detects signal from a large VOI
Depth resolved surface coil spectroscopy (DRESS)	$\mathbf{B}_0$ slice selection gradient	One-dimensional technique. it detects signal from slices parallel to the plane of a surface coil
Fast rotating gradient spectroscopy (FROGS)	$\mathbf{B}_0$ slice selection gradient/ selective presaturation	It detects signal from a large VOI

REFERENCES

1. D. I. Hoult: Rotating frame zeugmatography, *J. Magn. Reson.* **33**, 183–197, 1979.

2. S. J. Cox and P. Styles: Toward biochemical imaging, *J. Magn. Reson.* **40**, 209–212, 1980.

3. M. R. Bendall and R. E. Gordon: Depth and refocusing pulses designed for multipulse NMR with surface coils, *J. Magn. Reson.* **53**, 365–385, 1983.

4. M. R. Bendall: Elimination of high-flux signals near surface coils and field gradient sample localization using Depth pulses, *J. Magn. Reson.* **59**, 406–429, 1984.

5. R. E. Gordon, P. E. Hanley, D. Shaw, D. G. Gadian, G. K. Radda, P. Styles, P. J. Bore, and L. Chan: Localization of metabolites in animals using ^{31}P topical magnetic resonance, *Nature* **287**, 736–738, 1980.

6. R. E. Gordon, P. E. Hanley, and D. Shaw: Topical magnetic resonance, *Prog. Nuclear Magn. Reson. Spectroscopy* **15**, 1–47, 1982.

7. P. A. Bottomley, T. B. Foster, and R. D. Darrow: Depth-resolved surface coil spectroscopy (DRESS) for in vivo ^{1}H, ^{31}P, and ^{13}C NMR, *J. Magn. Reson.* **59**, 338–342, 1984.

8. P. A. Bottomley, L. S. Smith, W. M. Leue, and C. Charles: Slice-interleaved depth-resolved surface-coil spectroscopy (SLIT DRESS) for rapid ^{31}P NMR in vivo, *J. Magn. Reson.* **64**, 347–351, 1985.

9. R. Sauter, S. Mueller, and H. Weber: Localization in in vivo ^{31}P NMR spectroscopy by combining surface coils and slice selective saturation, *J. Magn. Reson.* **75**, 167–173, 1987.

APPENDIX B

Molecular Structure of Selected Metabolites[1]

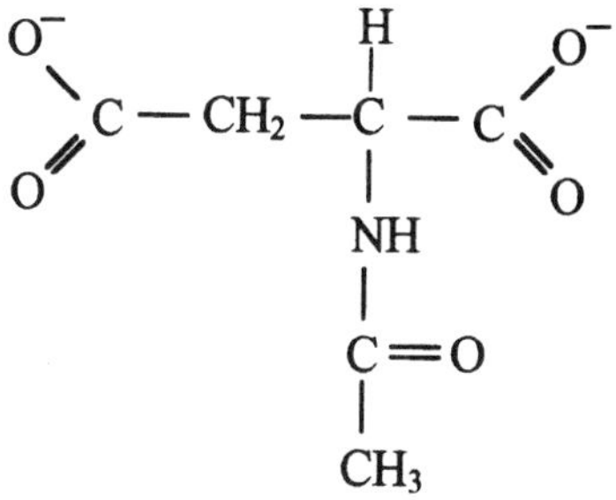

N-Acetyl aspartate

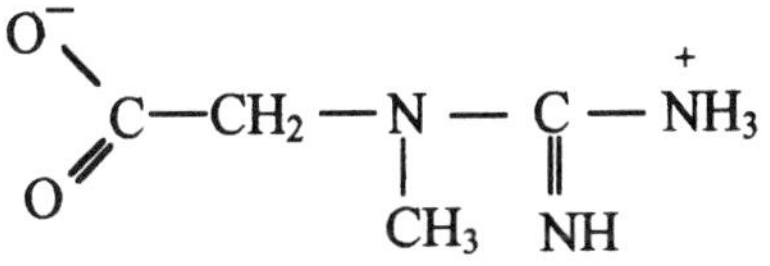

Creatine (Cr)

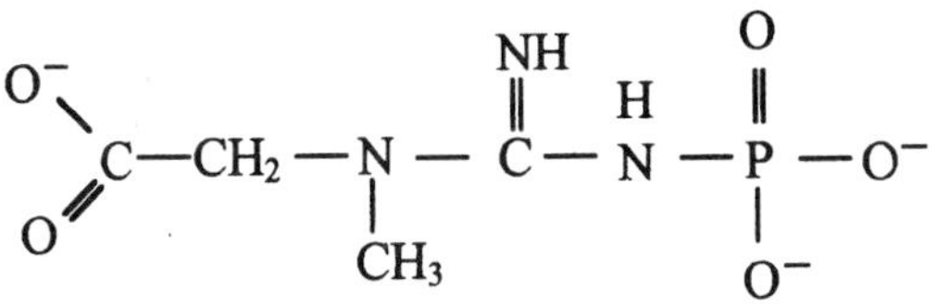

Phosphocreatine (PCr)

[1]Additional molecular structures are shown in Fig. 2.16

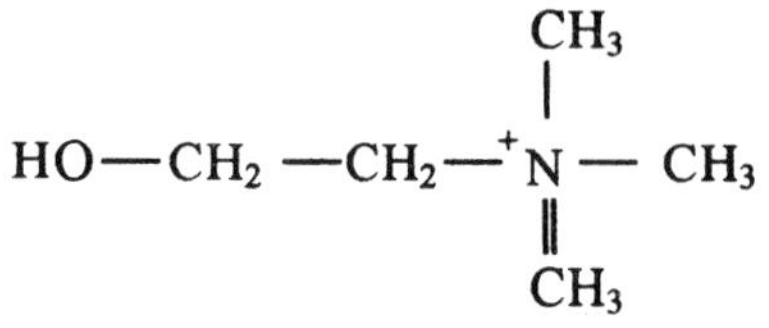

$$HO-CH_2-CH_2-\overset{+}{\underset{\underset{CH_3}{\|}}{\overset{CH_3}{\underset{|}{N}}}}-CH_3$$

Choline (Cho)

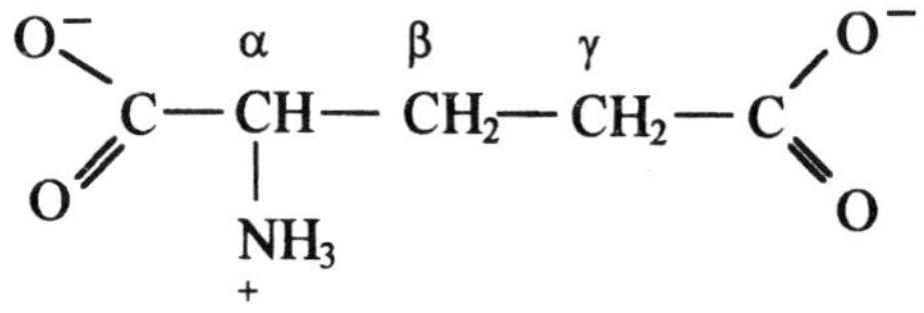

$$\overset{O^-}{\underset{O}{\diagdown}}C-\underset{\underset{+}{\underset{NH_3}{|}}}{\overset{\alpha}{CH}}-\overset{\beta}{CH_2}-\overset{\gamma}{CH_2}-C\overset{\diagup O^-}{\diagdown O}$$

Glutamate (Glu)

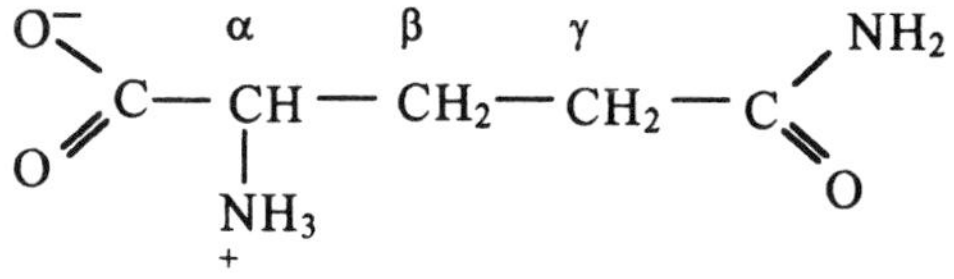

$$\overset{O^-}{\underset{O}{\diagdown}}C-\underset{\underset{+}{\underset{NH_3}{|}}}{\overset{\alpha}{CH}}-\overset{\beta}{CH_2}-\overset{\gamma}{CH_2}-C\overset{\diagup NH_2}{\diagdown O}$$

Glutamine (Gln)

$$\overset{O^-}{\underset{O}{\diagdown}}C-\overset{\alpha}{CH_2}-\overset{\beta}{CH_2}-\overset{\gamma}{CH_2}-\underset{+}{NH_3}$$

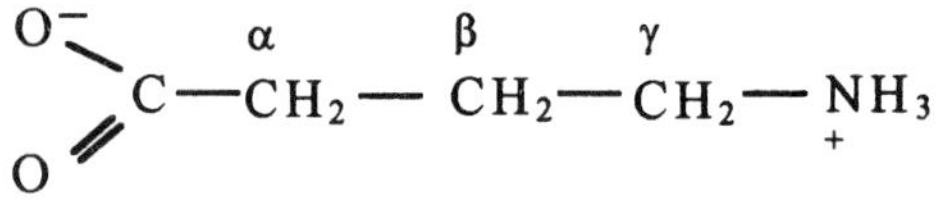

γ- Aminobutyric acid (GABA)

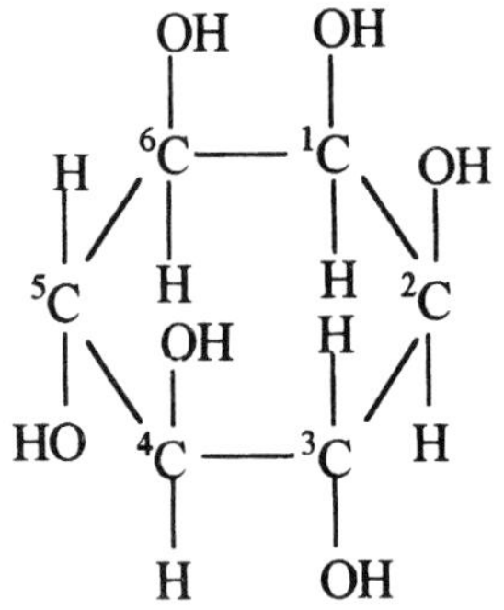

Myoinositol (mI)

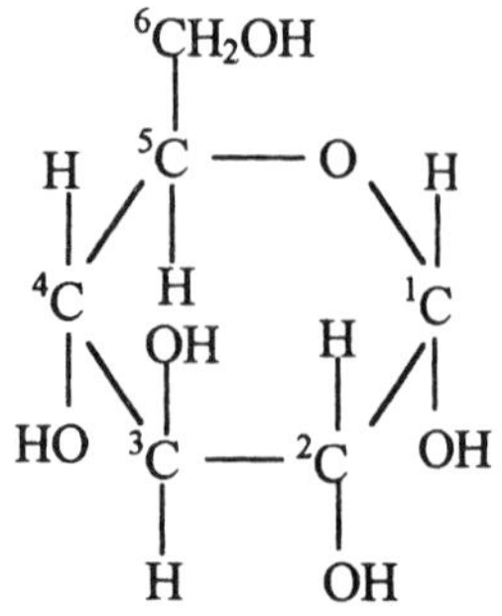

α-D- Glucose (Glc)

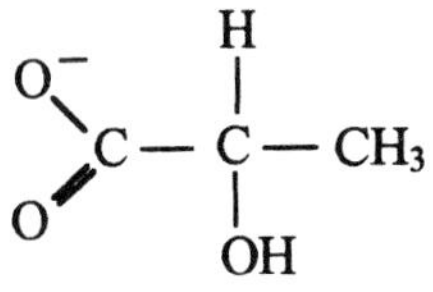

Lactate (Lac)

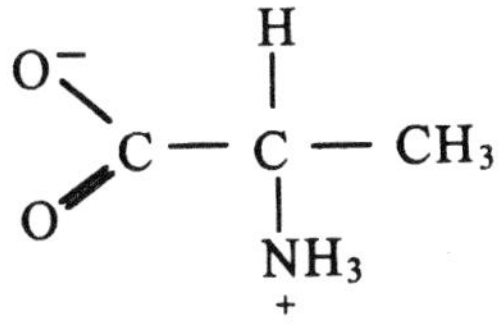

Alanine (Ala)

In this index, page numbers followed by the letter "f" designate figures; page numbers followed by "t" designate tables.